The Theory
of
Constraints

QUALITY AND RELIABILITY

A Series Edited by

EDWARD G. SCHILLING
Coordinating Editor
Center for Quality and Applied Statistics
Rochester Institute of Technology
Rochester, New York

RICHARD S. BINGHAM, JR.
Associate Editor for
Quality Management
Consultant
Brooksville, Florida

LARRY RABINOWITZ
Associate Editor for
Statistical Methods
College of William and Mary
Williamsburg, Virginia

THOMAS WITT
Associate Editor for
Statistical Quality Control
Rochester Institute of Technology
Rochester, New York

1. Designing for Minimal Maintenance Expense: The Practical Application of Reliability and Maintainability, *Marvin A. Moss*
2. Quality Control for Profit: Second Edition, Revised and Expanded, *Ronald H. Lester, Norbert L. Enrick, and Harry E. Mottley, Jr.*
3. QCPAC: Statistical Quality Control on the IBM PC, *Steven M. Zimmerman and Leo M. Conrad*
4. Quality by Experimental Design, *Thomas B. Barker*
5. Applications of Quality Control in the Service Industry, *A. C. Rosander*
6. Integrated Product Testing and Evaluating: A Systems Approach to Improve Reliability and Quality, Revised Edition, *Harold L. Gilmore and Herbert C. Schwartz*
7. Quality Management Handbook, *edited by Loren Walsh, Ralph Wurster, and Raymond J. Kimber*

ADDITIONAL VOLUMES IN PREPARATION

The Theory
of
Constraints

Applications in Quality and Manufacturing

**Second Edition,
Revised and Expanded**

Robert E. Stein
J.D. Edwards
Dallas, Texas

MARCEL DEKKER, INC.   NEW YORK · BASEL · HONG KONG

Library of Congress Cataloging-in-Publication Data

Stein, Robert E.
    The theory of constraints: applications in quality and
manufacturing / Robert E. Stein. — 2nd ed., rev, and expanded.
        p.   cm. — (Quality and reliability ; 50)
    Includes bibliographical references and index.
    ISBN 0-8247-0064-3 (alk. paper)
    1. Production management. 2. Reengineering (Management)
3. Computer integrated manufacturing systems. 4. Constraints
(Artificial intelligence) 5. Total quality management.   I. Title.
II. Series.
TS155.S7693   1997
658.5—dc21

                                                                96-52044
                                                                CIP

The first edition of this volume was published as *The Next Phase of Quality Management: TQM and the Focus on Profitability*, Marcel Dekker, Inc., 1994.

The publisher offers discounts on this book when ordered in bulk quantities. For more information, write to Special Sales/Professional Marketing at the address below.

This book is printed on acid-free paper.

MARCEL DEKKER, INC.
270 Madison Avenue, New York, New York 10016

Current printing (last digit):
10 9 8 7 6 5 4 3 2 1

PRINTED IN THE UNITED STATES OF AMERICA

To my wife, Debera, for her love and support;
to my daughters, Jennifer, Allyson and Elizabeth;
and to my granddaughter, Jessica.

About the Series

The genesis of modern methods of quality and reliability will be found in a sample memo dated May 16, 1924, in which Walter A. Shewhart proposed the control chart for the analysis of inspection data. This led to a broadening of the concept of inspection from emphasis on detection and correction of defective material to control of quality through analysis and prevention of quality problems. Subsequent concern for product performance in the hands of the user stimulated development of the systems and techniques of reliability. Emphasis on the consumer as the ultimate judge of quality serves as the catalyst to bring about the integration of the methodology of quality with that of reliability. Thus, the innovations that came out of the control chart spawned a philosophy of control of quality and reliability that has come to include not only the methodology of the statistical sciences and engineering, but also the use of appropriate management methods together with various motivational procedures in a concerted effort dedicated to quality improvement.

This series is intended to provide a vehicle to foster interaction of the elements of the modern approach to quality, including statistical applications, quality and reliability engineering, management and motivational aspects. It is a forum in which the subject matter of these various areas can be brought together to allow for effective integration of appropriate techniques. This will promote the true benefit of each, which can be achieved only through their interaction. In this sense, the whole of quality and reliability is greater than the sum of its parts, as each element augments the others.

The contributors to this series have been encouraged to discuss fundamental concepts as well as methodology, technology and procedures at the leading edge of the discipline. Thus, new concepts are placed in proper perspective in these evolving disciplines. The series is intended for those in manufacturing, engineering, and marketing and management, as well as the consuming public, all of whom have an interest and stake in the products and services that are the lifeblood of the economic system.

The modern approach to quality and reliability concerns excellence: excellence when the product is designed, excellence when the product is made, excellence as the product is used and excellence throughout its lifetime. But excellence does not result without effort, and products and services of superior quality and reliability require an appropriate combination of statistical, engineering, management and motivational effort. This effort can be directed for maximum benefit only in light of timely knowledge of approaches and methods that have been developed and are available in these areas of expertise. Within the volumes of this series, the reader will find the means to create, control, correct and improve quality and reliability in ways that are cost effective, that enhance productivity and that create a motivational atmosphere that is harmonious and constructive. It is dedicated to that end and to the readers whose study of quality and reliability will lead to greater understanding of their products, their processes, their workplaces and themselves.

Edward G. Schilling

Preface

Since the publication of the first edition in 1994, many changes have occurred in the acceptance of the Theory of Constraints (TOC) by mainstream industry. Mostly because of the efforts of the Avraham Y. Goldratt Institute (AGI) and its associates, many companies have begun to participate worldwide. There are numerous signs that TOC has gained a major foothold:

- The Constraints Management Special Interest Group (CMSIG) was created by the American Production and Inventory Control Society (APICS). CMSIG's primary goal is to support education in the Theory of Constraints. Two major seminars and a number of workshops were given in 1995 and 1996, and more educational seminars are planned for 1997. Discussions of TOC certification have also been held.
- The number of publications on the topic of TOC has increased dramatically. The list includes books on measurement systems, quality, information systems and the TOC thinking process (TP).
- The number of companies implementing TOC continues to increase and includes major leaders in almost every industry such as General Motors, Ford, 3M, Rockwell, Texas Instruments, Johnson Controls, United Airlines, U.S. Air Force, Morton International Automotive, Parr Instruments, EMC Corporation, Ketema A&E and Delta Airlines.
- Major software companies have begun creating alliances with constraint-based software vendors. Some have begun their own in-house TOC systems development programs.
- In 1995 the Automotive Industry Action Group (AIAG), an organization whose function is to create standards within the automotive industry, adopted requirements for vendors to implement the TOC concepts in its own operations.

The first edition was designed to explain the impact of TOC on Total Quality Management (TQM). This second edition has been expanded in five major areas to include

- Emphasis on policy analysis
- The use of the TOC-based information system in the support of quality manufacturing
- A discussion of supply-chain management
- A TOC glossary of terms
- End-of-chapter questions

Since the first edition, dramatic changes have occurred. This is reflected in the extensive revisions to Chapter 4 and the addition of two chapters that address the TOC-based information system and its impact on TQM as well as the effect of TOC on supply-chain management. To help in the education process, a TOC glossary of terms and end-of-chapter questions have been provided.

The objective in writing this book is to create an improved platform within the TQM umbrella for supporting the profit motive. As in the first edition, this book presents and expands the body of knowledge on TOC pioneered by Dr. Eliyahu M. Goldratt. It includes an augmentation of the TQM methodology and outlines the process of continuous profit improvement sorely needed by today's hard-pressed manufacturing companies. It also presents new tools for managing and focusing change that are dedicated to increasing profitability and addresses the implementation of the traditional tools from a global perspective.

. . .

The concepts of the Theory of Constraints have been proven to accelerate dramatically the profitability of any company.

. . .

I would like to offer my sincere thanks to some special people. Without their education, patience and dedication to a worthy cause, this book would not have been possible. Thank you, Dr. Eli Goldratt, Messrs. Avraham Mordock and Eli Schragenheim, and the staff and my former associates of the Goldratt Institute.

Robert E. Stein

Contents

1
Introduction

. . .

For us to merely duplicate the efforts of our most worthy opponent is to condemn ourselves to mediocrity. It is time we recognized that we can improve on the current level of understanding.

. . .

This chapter introduces the Theory of Constraints (TOC) and its application in a manufacturing environment and lays the groundwork for later chapters.

OBJECTIVES

- To introduce the Theory of Constraints
- To provide an overview for the rest of the book

THE CONCEPT OF THE THEORY OF CONSTRAINTS

The TOC, which is based on the natural laws that govern every environment, seeks to determine the underlying cause(s) of problems and to find the best solutions. This is its primary concern. It is *not* limited to dealing with the invalid policies or physical limitations within a manufacturing company, but has evolved to be applied to a variety of such diverse fields as

- Hospital management
- Engineering
- Quality management
- Information systems
- Military intelligence

- Sales and marketing
- Project management
- Distribution

It is virtually unlimited in scope and can be used in almost any field.

The cornerstone of TOC is the TOC thinking process. As explained in Chapter 4, the TOC thinking process focuses on three major issues:

- What to change
- What to change to
- How to accomplish the change

To accomplish a change successfully, it is necessary first to understand the underlying cause of a specific effect and then to determine what to add, subtract or modify to eliminate the effect. Just because the need for a change is obvious does not guarantee that it will be accomplished successfully. It is crucial to know how to accomplish the change without being blocked and without causing more harm than good.

THE IMPACT ON MANUFACTURING

In the mid- to late 1970s, Eliyahu Goldratt approached the task of solving problems related to the manufacturing environment. His approach, unlike that of many of his predecessors and peers, was to begin by gaining insight into the cause-and-effect relationships between the goal of the company, which in most cases is to make money, and the day-to-day decisions and actions of management and employees.

Understanding that most companies attempt to make more money now and in the future, Goldratt found that the actions taken to improve the profitability of a company often make no sense from a global perspective. In other words, there is often no connection—direct or indirect—between the actions of management and employees on one hand and the improvement of profitability on the other. This may seem dubious until one realizes that most of the processes being used in industry to manage profitability were designed at the end of the last century and that little has changed. Even when considering such relatively new movements as just-in-time (JIT) and total quality management (TQM), the biggest problem to overcome has been eliminating the pervasive influence of cost accounting.

As will be seen throughout this book, TOC represents a tremendous change in direction for most manufacturing companies. It introduces fundamental principles upon which to build a profitable foundation for any company, regardless of industry, including

- A new measuring system
- A process of continuous process improvement
- A fundamental decision process focusing on global rather than local issues
- A new method for analyzing the relationships between resources and determining where to focus efforts
- New insights into how to use the traditional TQM tools to maximize profitability
- New methods for scheduling the factory that have been proven to be superior to JIT
- New methods for analyzing policy problems and arriving at simple solutions

OVERVIEW

The scope of this book presents the impact of TOC in a number of areas within the manufacturing environment, including

- Quality
- Information systems
- Scheduling
- Decision making
- Product/process design
- Statistical analysis
- Supply chain management
- Strategic planning

The objective is to create a solution to the traditional approach of managing a manufacturing company so that it is more in line with the goal of the company while providing a logical framework for the overall structure.

To provide a proper foundation, the initial thrust of the book is devoted to changing the paradigms associated with traditional management strategies and then to offer a logical solution as well as a platform upon which to build the new structure. The discussion provides guidelines for implementation and insights into the strategic issues.

Chapters 1–6 represent a logical progression from absolute measurements used to guide the process of continuous profit improvement to the creation of an enhanced method of shop floor management. These chapters discuss the impact of establishing valid policy systems and offer an improved method for addressing the physical environment.

Once the basis for change has been developed, Chapters 7 and 8 provide a platform for implementing an augmented for of Total Quality Management, explaining the TOC/TQM connection and outlining organizational and communications strategies. It is one thing to say that something must be done but quite another to provide a global picture and to demonstrate *who* is responsible and *how* it can be accomplished.

Chapters 9–12 discuss well-established traditional tools such as quality function deployment (QFD), design for manufacturability, statistical process control (SPC) and design of experiments (DOE) from a traditional as well as a global perspective and illustrate what modifications must be made to take advantage of these techniques.

Chapter 13 introduces the concept of the TOC-compatible information system and the advantages provided in the support of the decision processes discussed in Chapter 5, "Correcting the Decision Process," while Chapter 14 presents TOC concepts in supply-chain management.

Chapters 15 and 16 discuss concepts of implementation and strategy. Once the reader is armed with full knowledge of what must be done, the task of how to accomplish and then fine-tune the overall strategy must be addressed.

CONCLUSION

What would be the impact of a program that could systematically identify those things that, if improved, would result in an immediate increase in profit and, if placed "end-to-end" would create a process of continuous profit improvement? This book is an attempt to document such a system by showing how to successsfully implement TOC for the manufacturing environment.

2

Creating the Process of Continuous Profit Improvement

• • •

What is needed is a method to increase the probability that when an action is taken to improve, it will have a positive impact on profitability as well as the overall health of the company.

• • •

In this chapter a new set of measurements designed to facilitate the improvement process as well as the process of continuous profit improvement are created.

OBJECTIVES

- To establish absolute measurements from which to judge the impact of any improvement process
- To determine where to focus overall improvement objectives
- To define the process of continuous profit improvement
- To gain insight on what type of problems will prevent improvement in profitability.

ESTABLISHING NEW MEASUREMENTS

Before any discussion of improvement can be entertained, a method of measurement that can be agreed upon by all parties must be identified. To say an improvement has occurred requires validation. Most would agree that return on investment (ROI) is an adequate measurement for determining whether an improvement has occurred. However, it does not provide insight into where to focus to produce improvements.

Owing to an increase in competition over the past 10 years, additional measurements have been developed for factors that can increase a company's competitiveness. These key indicators are the competitive-edge issues of quality, lead time and price. It was easy to develop a logical approach that would maximize attention to each of these issues so that efforts could be made to eliminate those causes that affect these factors negatively. This approach fell into place with the current concept of cost/waste reduction. However, it has become apparent that improving a company's position with respect to the competitive-edge issues may only serve to increase the *potential* for making more money. Such efforts do not guarantee increased profitability, and when approached without benefit of a thorough understanding of the impact of the total corporate environment, they may result in a much less competitive entity.

What is needed is a new set of measurements and a process of improvement that, when implemented correctly, will guarantee that profitability will go up. If this new set of measurements can be agreed on, then what is left is to determine the relationship between these measurements and the corporate environment. A company's decision processes and actions with regard to the process of ongoing improvement should follow naturally from there.

For a decision system to work, there must be a direct correlation between the measurements used externally, such as net profit (NP) and ROI, and those used internally. This is obvious. What is being proposed and gaining more and more acceptance as the measurements of choice are:

```
Throughput (T)       - The  rate  at  which  the  system  generates
                       money through sales.

Inventory (I)        - All  the  money  invested  in  purchasing  the
                       things the system intends to sell.

Operating (O/E)      - All the money the system spends in turning
Expense                Inventory into Throughput.
```

Throughput is represented by the formula Sales – Raw Material. Inventory includes any physical inventories such as work in process, finished goods and raw material but also includes tools, buildings, capital equipment and furnishings. Operating Expense includes expenditures such as direct and indirect labor, supplies, outside contractors and interest payments.

These measurements were first introduced by Eli Goldratt in *The Goal* (1984), *The Race* (1986) and *The Haystack Syndrome* (1990) and are easily understood by most people without hesitation. The objective is to maximize Throughput while minimizing Inventory and Operating Expense. The measurements relate directly to the way in which global goal attainment is measured (Goldratt, 1990).

```
Net Profit =  Throughput - Operating Expense

Return on Investment = Throughput - Operating Expense
                                  Inventory
```

They can be used to describe any number of additional measurements, which may give insight to their importance as a decision tool.

```
Productivity =      Throughput
                 Operating Expense

Inventory turns = Throughput
                   Inventory
```

Once established, these measurements can be used to analyze the impact of each internal decision on the external measurements of NP and ROI. Whenever a decision is made, the question becomes whether or not Throughput will go up or Inventory and Operating Expense will go down. Typical decisions include: where and how to focus process improvement efforts, what products should be sold and for what price, whether a setup should be torn down to run a hot order and, if so, how much should the customer be charged? Each of these decisions can be analyzed based on its impact on the key measurements of Throughput, Inventory and Operating Expense. If rework is reduced in a specific work center, will Throughput go up? If so, by how much? Will Inventory and Operating Expense go down? If so, by how much?

What is needed next is an understanding of the environment dictating the effect on each measurement.   In short, what are the occurrences in our environment that affect the creation of Throughput and the existence of Inventory and Operating Expense?

DECIDING WHERE TO FOCUS

Most people would agree, intuitively, if given only one choice of what to improve in the three measurements, they should choose Throughput as the most important. However, beyond simple intuition are some very convincing arguments.

- Inventory and Operating Expense are limited in their ability to be improved, since they cannot be reduced past zero and a certain amount of each is required to produce and to protect Throughput.
- A process of continuous profit improvement based on a program that concentrates on a reduction of Operating Expense seems rather unlikely, as it would be difficult to maintain. The closer to the objective of zero Operating Expense, the more difficult it becomes to continue. If the objective were actually met, then sales would also be reduced to zero.
- If Inventory and Operating Expense exist to produce and protect the Throughput figure, understanding more about what is to be done to improve Throughput becomes a prerequisite to dealing with the other two measurements.
- Throughput is not inherently limited and therefore produces the greatest opportunity for improvement. A tremendous number of Inventory and Operating Expense "sins" can be forgiven if the Throughput figure continuously grows at a fast enough pace.
- Concentrating on Throughput provides a tremendous leveraging effect in that since it is created by an interdependent sequence of events; very few things must be improved for Throughput to increase.

The 99/1 Rule

Whenever the demand for two resources in a chain of events has reached 100% capacity, as in the following illustration, the probability is very high that sometimes the first resource will not be able to deliver to the second.

```
        100%  100%  Load
    *----*----*----*
        25    25    Demand in Units
```

If this occurs, then the second resource will not be able to deliver its demand either. However, capacity is usually described as an average capability distributed along a bell curve. The probability is very high that the first resource will constantly deliver less than 25 units. Since this is the case, the probability is also very high that the second resource will rarely be loaded at 100% capacity. It will not have all the material necessary to "keep it busy."

· · ·

In any chain of events there can be only one weakest link, and if improvement is to occur only the weakest link needs to be strengthened.

· · ·

This is an advantage in determining where to focus. The 80/20 rule of the Pareto principle, by which 80% of the cost is created by 20% of the cost drivers, can be adapted as the 99/1 rule (Goldratt, 1990), where 99% of the impact results from 1% of the change.

The Impact on Corporate Functions

Since Throughput is described as the rate at which money enters the company, all functions must be described by a rate: for example, the rate at which parts are purchased, the rate at which designs are created or the rate at which products are sold. Any time a rate is established, capacity must also be considered: the capacity to buy parts, the capacity to design, the capacity to sell and so forth.

● ● ●

Not only the physical resources but also the individual functions of a corporation are subject to the laws governing probability and statistical fluctuation.

● ● ●

If the ability to deliver is limited by capacity in production, then an increase in the ability to sell will do nothing to increase the rate at which Throughput enters the company. This phenomenon offers a very distinct advantage in that very few things must improve in order for Throughput to go up.

THE NEED TO DEFINE THE PROCESS OF CONTINUOUS PROFIT IMPROVEMENT

A process of improvement should include a sequence of steps that should end with an improvement in a company's profitability. This process, when repeated, should produce a sequence of improvements. When repeated endlessly, it should produce a process of continuous profit improvement. The 14 points so thoughtfully provided by Dr. W. Edwards Deming were never meant to fulfill the requirement as a sequence of steps, but rather as a comprehensive structure for organizing the TQM program. These 14 points, which have brought about such tremendous change, are as follows:

1. Maintain a constancy of purpose
2. Adopt a new philosophy
3. Eliminate dependence on mass inspection
4. Cease price-alone purchasing

5. Plan constant improvement
6. Improve job training
7. Provide a higher level of supervision
8. Eliminate unsuitable material
9. Drive out fear by encouraging two-way communication
10. Get rid of numerical goals and slogans
11. Examine closely the impact of numerical standards
12. Teach and utilize statistical techniques
13. Institute a vigorous training program in new skills
14. Institutionalize the above points

Shewhart's concept of plan-do-check-action begs the obvious question, What should be planned? Of all the activities of a company, the ones that should be elevated to the highest priority should be those that pertain to making money. Unless a process of improvement can be defined that can predictably accomplish this, then attempted improvements at profitability will be a hit-or-miss proposition.

DEFINING THE PROCESS STEPS

A process of improvement for Throughput does not automatically assume that the location of the weak link is known. If something is acting to restrict the amount of Throughput being generated, then the first step should be to identify it. Once the weak link has been identified, steps must be taken to ensure that the amount of Throughput being generated is maximized. There are two issues:

- The amount of available capacity at the weak link to apply to generating Throughput
- The characteristics of the other resources that can be applied to ensure that the weak link is not restricted in any way.

Figure 2.1 represents a chain of events in which resource 1 is the gating operation. Each resource feeds the next until resource 4 is reached. The capacity of each resource is different and refers to the delivery capability in units. From this

```
Resource      1     2     3     4
              *----*----*----*
Capacity     70    60    25    40   =   25 Units
```

Figure 2.1 Chain of events.

figure it is easy to see that the output limitation for this line is 25 units. It is also easy to see that in order to increase Throughput, only resource 3's capability needs to be addressed. So, the improvement process must begin there.

Before simply buying another resource 3, it may be more profitable to see how much additional productive time can be gained from that resource as it currently exists. It may be found that resource 3's capability can be increased without buying a new resource. Maximizing the amount of productive time at the weak link may include:

- Creating a schedule for the weak link which uses every available minute of time effectively
- Selling a product mix into the market which maximizes the amount of Throughput generated
- Reducing the amount of setup and maintenance time to maximize resource availability

Once a strategy for delivering the maximum amount of Throughput from resource 3 has been developed, this strategy must be protected. It is the other resources within the system and their characteristics that supply this capability. If resource 2, for whatever reason, does not deliver all of resource 3's parts requirements, this will limit the number of units being created. If resource 3 is an NC machine and needs programs to run but the engineers are not delivering on time, Throughput would suffer. On the other hand, if all resources were to work at 100% efficiency and utilization the amount of Inventory in front of resources 2 and 3 will begin to climb. The result would be that return on investment would decrease, lead times would be expanded and, eventually, Inventory would block the creation of Throughput. So, all nonconstraint resources, including functional organizations such as sales, engineering and quality, should deliver to the constraint what is needed for it to maximize the creation of Throughput, and nothing more.

After maximizing the capability of resource 3 and ensuring that all other resources are delivering to resource 3 what it needs, another strategy must be adopted to increase the amount of Throughput generated—the addition of another resource 3.

As shown below, resource 3 has been elevated by purchasing another machine to doubling its capability. The assembly line is now capable of producing at the rate of 40 units.

```
Resource    1     2     3     4
            *----*----*----*
Capacity   70    60    50    40   =   40 Units
```

Notice that when this occurs, the weak link has moved and is now in another location. It becomes obvious for a process of continuous improvement that Throughput must now be maximized for a different resource and that all other resources must be aware of those requirements for protecting the new weak link. To make a further improvement, capacity at resource 4 must then be increased. Because of this change, the company may need to reassess its approach as far as where to focus on new improvements and how to support the new weak link. This possibility must be considered each time a weak link has been elevated.

The process of improvement for Throughput as defined by Goldratt (1990) is:

- Identify the weak link, or *constraint*
- Determine how to maximize its capability to produce Throughput, or *exploit* it
- Determine what to do with all other resources to support the constraint, or *subordinate*
- Strengthen the weak link, or *elevate*.
- *Repeat* the process.

Note that it is important to remember that when the weak link has been strengthened, there is a tendency to continue to use the same solutions even though the weak link has moved and the solutions have become obsolete—avoid *inertia* (Goldratt, 1990).

. . .

The process of improving Throughput is not to be confused with the Shewhart's plan-do-check-action. In any chain of events, only one thing must be fixed to increase Throughput.

. . .

IDENTIFYING THE WEAK LINK(S)

Whenever decisions are made from a global perspective, the alternatives are determined by the limitations of the system and not by an isolated, localized algorithm. To know where to focus to make an improvement requires that the limitations and their impact must be known. This raises the question of what must be done to determine the limitations of the system?

By definition, unless the company is making an infinite amount of money, there must be a weak link. How should it be identified and what are the elements that limit the weak link's ability to perform to its maximum? Weak links are categorized in a number of different ways: behavioral, managerial, capacity, market, logistical, and managerial, each having its own impact on the smooth operation of the company. Logistical constraints involve limitations placed on the system by the planning and control systems. Managerial constraints are erroneous management strategies, policies and decision mechanisms. Behavioral constraints are those behaviors and work habits exhibited by employees which result in poor performance from a global perspective.

Behavioral Constraints

Behavior is the result of an attempt to act or react, in a logical way, to the environment and specific situations encountered. It is directly affected by the training, education, measurement systems, experiences, attitudes and mental dispositions of the people involved. Whenever a behavior is in conflict with reality and results in a negative impact on the global measurements of the company, it is said to be a behavioral constraint. Behavioral constraints are caused by a number of different reasons. Probably the most prevalent cause is linked to the measurement system. "Tell me how you will measure me, I will tell you how I will behave" (Goldratt, 1990). Whether implicit or explicit, the measurement systems dictates the way that people act. The best example of this is the concept of staying busy.

One of the hardest behaviors to change—and yet it may be one of the most devastating—is the concept that resources must stay busy. The assumption is made that whatever an employee does to "stay busy" will result in good things happening. It is reinforced by the utilization measurement, whereby every resource must be highly utilized or else the company will lose money. This concept is held by management and employees alike, although not necessarily for the same reasons. The extended result of this kind of behavior is that inventories begin to climb, product mixes become unbalanced, schedules slip and material shortages occur.

Another example of a behavioral constraint is in the tendency to maximize savings during setup. Planning setup this way, without knowledge of the global impact on Throughput, Inventory and Operating Expense, may result in a decline in profitability. When viewed from a global perspective, this approach may seem almost irrational, and the negative impact on profitability is often predictable. And yet, it is very difficult to convince a foreman to act otherwise.

Managerial Constraints

Poor management policies often act to restrict the ability to maximize the utilization of physical resources or to prevent the proper use of nonconstraint resources in protecting the creation of Throughput. As an·example, a policy of setting commission schedules for sales representatives using activity-based accounting to determine which products to push onto the market may cause the poor exploitation of resources for maximizing profitability. Such a policy may, in fact, cause a serious decline in profitability. Or, a policy of establishing quality cost as the mechanism for focusing improvement may result in money spent to improve an area that will not help to increase the overall profitability of the company. (See Chapter 5, "Correcting the Decision Process.") *We are a defense contractor* is the sort of common notion that may impose severe market limitations. The only way to change this may be to change the mind of the board of directors.

Capacity Constraints

A capacity constraint exists any time the demand placed on a resource exceeds its available capacity. Capacity constraints can include machines or people and can restrict the creation of Throughput. Primary constraints are those that restrict the output of the entire company. Secondary constraints restrict the ability to properly subordinate to the primary constraint. In other words, if the demand placed on a resource increases to the point where the probability is low that it will be able to deliver to the primary constraint what is needed, the problem is said to be a secondary capacity constraint.

Market Constraints

Perhaps the most important constraints to consider are those created by the market. The market controls the product, pricing, lead time, quantity and quality of the goods and services demanded, and it establishes the necessary conditions for creating Throughput. Whenever market demand is less than the capability of the company's resources, a market constraint exists. While market constraints have many causes, most exist due to management policies.

Logistical Constraints

Anytime problems occur that originate from the planning and control systems within the company, there is said to be a logistical constraint. Material requirements planning systems that are capacity insensitive create problems in the proper synchronization of resources and can escalate the amount of Inventory and production problems that already exist. For example, a cumbersome purchasing process in which for every purchase the lowest price must be selected from a minimum of three bids from three different vendors for every purchase may actually restrict the creation of Throughput.

Necessary Conditions

A growing number of companies are being required by their customers to implement statistical process controls. If this demand is not met, they will be discontinued as approved vendors. Necessary conditions are boundaries or demands placed on companies, departments or individuals, originating internally or externally, which serve to regulate activity. They may include government regulations, such as environmental restrictions involving the disposal of toxic waste, customer demands or moral issues, such as honesty. Management may place necessary conditions on employees. Stockholders place necessary conditions on companies for the purchase of stock. Whenever a necessary condition is not being met, there are usually serious implications. If a necessary condition of employment is honesty and the employee is caught stealing, then he or she will probably be fired. If a necessary condition for being in business is adequate cash flow and that condition is not present, then the company will probably go bankrupt. Necessary conditions should be distinguished from constraints. A necessary condition can become a constraint unless the condition is met. Once met, any improvement in it will not continue to improve the profit position of the company. However, it still must be continually enforced.

The Cost Mentality

Most of the irrational behavior exhibited in companies originates from a "cost mentality" (Goldratt, 1990). The cost mentality results in the tendency to optimize local measurements at the expense of global measurements. In most companies, there seems to be almost a natural tendency to departmentalize organizations and to develop measurements that optimize the results at a local level. The use of local measurements has been taught in the business schools for decades and further reinforced in the business world. However, unless an improvement in the local measurements is supported by a positive global impact, then the local measurement system must be viewed as invalid. The cost mentality is the basis for most constraints and is probably our biggest enemy.

STUDY QUESTIONS

1. What measurements are used to measure the success of a company from a global perspective?
2. What local measurements are suggested by the Theory of Constraints as being valid for supporting the local decision, and how are they reconciled with the global measurements?

3. Define Throughput as a local measurement. From what two measurements is it constructed, and how?
4. What is the 99/1 rule and what is its impact on the improvement process?
5. List at least four reasons/benefits for focusing improvement efforts on Throughput.
6. How is the goal of most corporations defined?
7. What are the steps of continuous profit improvement, and how do they differ from Shewhart's plan-do-check-action?
8. List and define the five types of constraints.
9. Define and give two examples of necessary conditions.
10. What is the concept of the cost mentality, and what is its overall impact on the conventional corporation? Give examples.

3

Dealing with the Physical Environment

Chapter 3 begins the process of explaining how to address problems associated with the physical environment and what the global impact of these problems is from a profitability perspective.

OBJECTIVES

- To understand how resources interface and their resulting impact on the system (product flow)
- To begin to understand the impact of physical limitations from a global perspective
- To explore the impact of physical limitations on currently accepted technology and how to change them.
- To understand how and where to focus physical improvements
- To introduce the concept of the diagnostic system and to understand the problems inherent in the ability of current information systems technology to identify physical limitations

IMPROVING IN THE DEPENDENT VARIABLE ENVIRONMENT

Any attempt to improve the performance of constrained resources must begin with the right question. In addressing the choices, it is important not to be misdirected by preconceived notions of what activities will be necessary. An improvement just for the sake of improving may actually reduce profitability. The question that should be asked is, How can the Throughput of the company be increased, given the current situation? Properly framed, the question can be properly answered.

In any chain of events, each link within the chain is capable of impacting any other link. It is the characteristics of the links and their relative position within the chain that dictate how each must interact with the others and that determine the overall effectiveness of the system.

One resource can negatively effect another by:

- Not delivering
- Delivering late
- Delivering poor quality
- Delivering incorrect products
- Delivering too early

Resource can show variation in these effects. A resource may sometimes deliver the right product on time and deliver a high-quality product as well. It is the characteristics of each resource that dictate its ability to cope with what it has received and that may improve or make worse the position of the next resource. A resource that has additional capacity to deal with a rework problem is in a much better position than a resource that has no additional capacity. Understanding this issue, while it may seem very simple, has had far-reaching impact on the overall TQM program in its ability to focus as well as manage improvement.

• • •

The dependent variable environment is the environment in which resources are dependent on each other in their capability to produce and are subject to variation in that capability.

• • •

UNDERSTANDING PRODUCT FLOW

Creating the Product Flow Diagram

While constraints may come in many different flavors, to begin to understand the relationships between resources, and therefore to predict the specific impact each resource will have on the overall system, it is necessary to understand how products flow through the factory. The product flow diagram is a detailed description of the product flow and the resources involved in the production operation. It is a cornerstone to organizing any TQM program and is somewhat different from the process flow diagram in that primary interest is placed on determining the sequence of operations and the resources that are used. Storage and transportation are ignored.

The basic building block of the product flow diagram is the part/operation, or station. The station is where an operation is performed on a specific part. The

Indented Bill of Material		Routing Part	Op.	Res.
123	Make	123	10	R-4
124	**Make**		20	R-5
125	Purch		30	R-6
126	Make	**124**	**10**	**R-3**
127	Purch		20	R-2
			30	R-1
		126	10	R-3
			20	R-2
			30	R-1

Stations List		Resource Load	
123/30	R-6	R-1	100%
123/20	R-5	R-2	75%
123/10	R-4	R-3	50%
124/30	R-1	R-4	70%
124/20	R-2	R-5	90%
124/10	**R-3**	R-6	60%
125/Purch			
126/30	R-1		
126/20	R-2		
126/10	R-3		
127/Purch			

Figure 3.1 The basic building blocks of the product flow diagram.

correct sequencing of stations plus the addition of the resource upon which the operation is performed create the product flow diagram. In Figure 3.1, the indented bill of material shows which parts are used on which assembly operations. The routing shows the operations that must be performed, the sequencing and the resources involved for each part. The stations list shows all the different stations that are created by combining the bill of material and routing file into one structure. Also included is resource load information.

Notice that in the routing, part number 124's first operation is operation 10 on resource R-3 (in bold). The station in the stations list is therefore 124/10 (in bold).

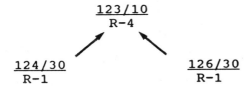

Figure 3.2 Convergent operations.

When creating a visual diagram, one combines stations by using arrows to define the production flow. Arrows designate the direction of flow to and from specific stations. Stations that have more than one arrow converging are assembly operations or convergent operations (Figure 3.2).

Stations/resources 124/30/R-1 and 126/30/R-1, which are the last steps in creating parts 124 and 126, feed station/resource 123/10/R-4.

Stations that have multiple outgoing arrows are divergent operations (Figure 3.3). Material flowing from station 1234/10 diverges to stations 1245/30 and 1267/30. Converging and diverging operations cause unique problems under certain circumstances, as will be seen later in this chapter.

The product flow diagram illustrates the chain of events for the production process. In this way, one can understand what stations occur within the chain.

Based on the route file, the bill of material file and the resource load presented, the product flow diagram would look like Figure 3.4.

Notice that at the top of the structure, the last part/operation is being performed, and at the bottom, raw material enters. In this manner, it is easy to construct the chain of events occurring within the factory, from the beginning to the end, for each order created by the customer.

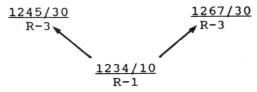

Figure 3.3 Divergent operations.

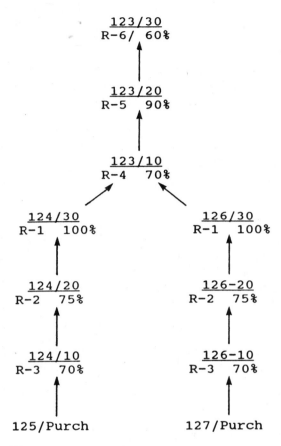

Figure 3.4 The product flow diagram.

The Impact of Capacity

Whenever the capacity/demand of each resource is calculated, a better picture is created for determining where to focus to maximize the global impact. Resources that have been loaded at or near capacity are particularly vulnerable. These resources should be given close attention during the improvement process. However, it is equally important, although not as obvious, to ensure that once a product has passed the constraint, it will not be scrapped or require rework, both of which absorb constraint time. The chain of resources that leads from stations 124/30 and 126/30, performed on resource R-1, to station 123/30, performed on resource R-6, must be very well protected from scrap problems. Rework problems occurring at station 123/10, performed on resource R-4, and loaded at 70% cause less of a problem than those rework problems occurring at station 123/20, performed on resource R-5, and loaded at 90%.

In the *process flow* diagram, emphasis is placed on increasing the efficiency of how resources interface or reduce waste. This is a highly questionable approach from a global perspective. In the *product flow* diagram, the issue is to increase or protect the creation of Throughput. A closer look at the impact of product flow is made in Chapter 11, on the implementation of Statistical Process Control.

IMPROVING PERFORMANCE OF CONSTRAINED RESOURCES

Techniques for increasing the capacity of constrained resources at a local level deal with the different categories of resource time, including processing time, setup time, and idle time. The objective is to convert setup and idle time to processing time and to minimize the processing time.

THE QUICK SETUP STRATEGY

The SMED Process

Originally conceived and documented by Dr. Shigeo Shingo, the Single-Minute Exchange of Die (SMED) strategy is a mandatory technique for any small-lot-size production strategy. Over a 19-year period Dr. Shingo created a technique that has drastically reduced the amount of setup time required in production by implementing a three-stage process of:

- Separating internal from external setup
- Converting internal to external setup
- Streamlining all aspects of the setup operation

Internal setup (IED) describes those operations that are performed while the process is stopped; external setup (EOD) describes those operations that are performed while the process is running.

The first stage in this process is to identify and separate those activities that are normally done while the process is stopped and those that are normally done while the process is running. A large amount of time during setup, as much as 30–50%, is usually spent on tasks that can be done while the machine is running but that, owing to poor organization, are often done while the machine is stopped, such as trying to find essential parts, jigs or fixtures. A specific tool may be needed to move a fixture onto or remove a fixture from a specific machine. This same tool may be used by other machines and operators in similar setup operations on numerous machines. However, unless the tool is at the proper location, at the proper time and in proper working order, it is useless. Stopping the constraint to hunt for the right fixture or tool can be disastrous. The loss is equal to the Throughput loss of the chain. If only one chain exists, then the Throughput of the entire corporation will suffer. A good first step to implement would be to ensure that everything needed to accomplish the setup is available at the machine when it is shut down. Checklists can be developed which itemize the requirements, function checks can be made to determine if fixtures are in working order and the movement of material and fixtures to and from machines can be streamlined and accomplished while the machine is running.

The second stage is to convert those activities that are normally done while the machine is down so that they can be done while the machine is running. If stopping the constraint will result in a great loss of Throughput, everything preventive action must be taken to minimize the amount of downtime. There are two approaches: (1) converting those activities that are assumed to be internal by re-examining their function and (2) identifying those activities that can be converted by a process change. For example, the utilization of dual-purpose dies, where one die may fit a number of different products, means that product changes can be accomplished without the need for stopping the operation.

The third stage is to further reduce the amount of setup by reducing the time required to perform the remaining internal tasks. In some cases this may include radical process changes; in others it may include small changes, such as shortening the amount of turns required on a bolt or replacing the bolt with a clamp. A large amount of time is spent studying the internal setup process for function. Individual steps are examined to determine whether they are required for the actual function that must be done and, if not, whether they can be eliminated or modified. As an example, in some machining operations, clamps are used that require a large number of bolts—as many as 15–20—which must be removed each time the clamp is removed. Each bolt may take a relatively long time to remove. In this case, shortening the bolt will not be a satisfactory solution because the strength of the clamp would be impaired by the use of shorter bolts. An examination of the function and assumptions made for each part of the clamping operation and a rethinking of the task to be performed reveals that removing the bolts is not be required after all. By cutting a U-shaped hole in the side of the washer and modifying the bolt holes in the clamp so that they will fit over the bolt heads, the washers can be removed after one turn of the bolt and the clamp removed over the top of the bolt heads. There are two major issues: (1) determining whether a function is required by examining the assumptions made and, if it is required, (2) determining whether a task can be modified to accomplish the required function, and how to modify it.

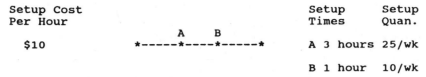

```
Setup Cost                                        Setup        Setup
Per Hour                                          Times        Quan.
                          A     B
   $10               *-----*----*-----*           A 3 hours 25/wk

                                                  B 1 hour   10/wk
```

Figure 3.5 Setup cost.

SMED Implementation

The method of SMED implementation varies from company to company and may include strategies to support small-lot-size operations, attempts to improve bottleneck operations so that production is increased, or efforts to reduce cost. With respect to cost, priority is usually given to those resources requiring the greatest amount of time to set up or is based on a cost of setup algorithm. However, the third principle of TQM II states that the value of an activity is determined by the limitations of the system (see Chapter 7). So, the value of a setup reduction activity from a global perspective is linked directly to the specific limitation being addressed. If the setup is being reduced on a non-constraining resource, regardless of the cost implications, the impact on the profitability of the company may actually be negative. Figure 3.5 illustrates this.

Resource A feeds resource B and has a setup time of 3 hours at a cost of $10 per hour, or $30 per setup. Resource B's cost is also $10 per hour but for only one hour, or $10 per setup. An engineer wants to spend $2,000 to eliminate the setup at resource A. Since resource A looses 75 hours per week to set up at $10 per hour, or $750, he estimates that he will be able to recapture the $2,000 expenditure in 2.7 weeks. However, Figure 3.6 adds some crucial information usually not considered.

Resource A is a non-constraint resource, with excess capacity loaded to 50%, while resource B is the constraint since it is loaded to 100%. Any reduction in setup time at resource A will only result in an increase in the amount of excess capacity available at that resource. The productivity model illustrates what is important.

```
Productivity =        Throughput       =    No Change
                   Operating Expense =      ^ $2,000
```

From a global perspective, Operating Expense is represented by money paid out to employees in the form of payroll, utilities such as water or electricity, rental for buildings or machines and employee benefits. It is highly unlikely that the expenses in any one of these categories will decline through the

```
        A      B
*-----*-----*-----*
      50%   100%
```

Figure 3.6 The impact of product flow on setup.

reduction of setup time. No one is going to be laid off; the plant is not going to be reduced in size and employee benefits will not decline. However, if the $2,000 check is written, Operating Expense will immediately increase by the $2,000. If Throughput does not go up, productivity as well as profitability will decline.

If the engineer had spent money to increase the available capacity at resource B (just the opposite of the cost model), although Operating Expense still would not have declined, the amount of Throughput entering the company would have gone up to offset the additional Operating Expense created by spending the $2,000.

As seen in the above example, implementing SMED to reduce cost is a poor focusing mechanism, which can lead to a loss in profit. But, by focusing on bottleneck/constraint operations, improvement can have an immediate positive effect.

· · ·

Bottlenecks occur any time the demand placed on a resource is equal to or more than capacity.

A constraint is anything that prevents the system from attaining its goal of making more money.

· · ·

Setup reduction programs should be planned to increase available time on the constraint resource(s) as part of the Exploitation phase, and near-constraint resources as part of the Subordination phase of the five-step improvement process.

Avoiding Inertia

Many times resources become constrained for reasons not related to setup or process issues. Poor scheduling or resource utilization can create bottlenecks where none should exist. A misinterpretation of data can lead to setup reduction programs that have no impact, or even a negative impact, on the bottom line. An electronics manufacturer of communications equipment that was operating under

Just-in-Time (JIT) was busily increasing the efficiency at bottleneck operations so that Throughput would increase. The capacity report had indicated that several resources were scheduled at or well above capacity (83 to 144%). However, no overtime was being worked and the due date performance to the customer was at 98%—a good physical indication that the company's assumption of a physical constraint existing in production was invalid. Upon closer examination, it was found that the real constraint to making more money was an inability to obtain sufficient raw material to release to the floor. A further examination found that purchasing was placing a high value on the reduced price of raw material and that products being received were being rejected at incoming inspection for being inadequate. Any time spent increasing the capacity of bottleneck resources at this facility would be a total waste. The only way of increasing Throughput at this facility at this time would be to change a policy in purchasing.

The resources that were considered bottlenecks may have been constraining Throughput at one time or another but were either improved or became non-constraining due to a new policy created in purchasing. It is very important to repeat the five-step process once the constraint has been broken. However, it is sometimes not so easy to see when to do it.

Creating Common Sense

The success of any SMED program has depends directly on the ability to examine the underlying assumptions made in any operation. While it is worthwhile to examine the actual technology employed by Dr. Shingo, and others, in the implementation of SMED, it may be a better idea—and in the long run more profitable—to examine the thought processes behind the creation of the solution. Common sense, it seems, has its birthplace in the ability to recognize the obvious from the not so obvious. What keeps people from recognizing those things that are obvious is that their thinking process is already shaped by preexisting assumptions before the process begins. In the five-step improvement process described earlier, inertia is caused by preexisting assumptions. The electronics company described earlier may have started with a legitimate resource constraint but failed to start the analytical process over again once the constraint was broken and had moved to purchasing. It is desirable for every worker, manager, and executive to have the ability to examine any situation without the negative impact of preexisting assumptions. This is the purpose of the current reality, evaporating cloud, future reality, prerequisite and transition tools described in Chapter 4, "Analyzing Policy Constraints." This is why Dr. Shingo suggests as part of his setup reduction process that the function of each step and tool in the process be examined and redefined. Breaking the underlying assumptions behind an activity or function gives rise to common sense. The new tools described in Chapter 4 add a much-needed dimension to the setup reduction program.

THE TOTAL PRODUCTIVE MAINTENANCE (TPM) STRATEGY

The TPM Process

The proper maintenance of equipment has a major impact on the overall profitability of any company. According to Seeichi Nakajima, in his book *Total Productive Maintenance* (Nakajima, 1988), TPM involves a companywide effort to create a "fundamental improvement within a company by improving worker and equipment utilization" while maintaining a low life cycle cost of maintenance. TPM is a system of maintenance for the entire life span of the equipment to include maintenance prevention (designing equipment to be maintenance free), preventive maintenance (ensuring that equipment remains in good working condition), corrective maintenance (repairing and reengineering broken equipment), and autonomous maintenance (operator involvement). Attempts are made to eliminate the "big six losses" of equipment failure, setup and adjustment, idling and minor stoppages, reduced speed, process defects and reduced yield. The effectiveness of the overall system is measured based on availability, efficiency and quality. Availability is determined by the following formula.

$$\text{Availability} = \frac{\text{Operation Time}}{\text{Available Time}}$$

Operation time refers to the net available time during the day minus equipment down time. Available time refers to the total time available in the day minus planned maintenance. Efficiency is determined by:

$$\text{Efficiency} = \text{Net Operating Rate} \times \text{Operating Speed Rate}$$

The operating speed rate is determined by dividing the theoretical cycle time by the actual cycle time. The net operating rate is determined by the actual processing time divided by the operation time.

Quality is described in TPM as a function of the defect rate and is quantified by the following formula.

$$\text{Defect Rate} = \frac{\text{Processed Amount} - \text{Defect Amount}}{\text{Processed Amount}} \times 100$$

The TPM effectiveness rating formula is:

$$\text{Effectiveness} = \text{Availability} \times \text{Efficiency} \times \text{Quality}$$

In traditional TPM, the maintenance activity is focused on the volume of occurrence. The resource with the lowest effectiveness rating is given the highest priority.

Although not mentioned in Nakajima's book, other indicators of system health include mean time between failures (MTBF) (the average time it takes for a machine to go from one failure to another), mean time to repair (MTTR) (the average time it takes to go from start to finish on a single repair), and the system's historical track record according to statistical process control (SPC), that is, whether the process can be controlled and how often it requires adjustment or repair. Under maintenance prevention, the focus should be on designing into the resources a high MTBF and low MTTR rating as well as quick setup strategies.

Changing the TPM Focus

Figure 3.7 shows how under the traditional TPM approach, resource C, with a 66% effectiveness rating, is obviously the problem resource needing immediate attention. However, from a global perspective, there are other things to consider. If resource A is the constraint and has an overall effectiveness rating of 82%, it could very well be a critical problem and is not considered by the current TPM focusing mechanism (Figure 3.8).

Resource A is loaded to 100%. Therefore, any loss of time at resource A results in a loss of Throughput, which may never be recaptured. Any scrap

Res	Avail.	Eff.	Qual. =	Effect Rating
A	90%	95%	97%	82%
B	90%	93%	95%	80%
*C	80%	92%	90%	66%

Figure 3.7 The TPM effectiveness rating.

```
        A     B     C              Resource
  *----*----*----*----*
       100%  90%   50%             Load
```

Figure 3.8 The impact of product flow on TPM.

results in a loss equal to the entire sales price. (The cost of replacing the raw material due to scrap plus the Throughput amount lost by having to recreate the product because of the constraint is equal to the sales price.) This means that 18% of total sales would be lost forever. Resource B, at 90% load, should definitely be considered the second most important potential problem. Resource C is loaded at only 50% and should be given a relatively lower priority than the other two resources unless the problems of resource C will negatively impact products already past resource A. The effectiveness rating, by combining availability and efficiency measures with quality, can mask the importance of the quality rating. Depending on where in the process a quality rating is measured, its level of importance may be magnified, regardless of the amount of excess load available. In Figure 3.7, if the quality rating is the recorded scrap rate for each resource, there will be a cumulative effect, resulting in the loss of 17% of total sales generated $[(.97 \times .95 \times .90) - 1]$. This masking effect may be the most important issue to address.

The health indicators of a machine, regardless of current demand, will give an indication of problems that may impact Throughput at a later time. Poor health indicators should never be ignored, even if the current impact does not appear significant. If a machine or group of machines that are loaded to 50% should suddenly break down for a period of time much longer than expected they could become temporary constraints and could block the Throughput of the entire company. Principle 5 states that the utilization of any resource may be determined by any other resource in a chain of events. Note that a poor machine rating is also an indicator that bad parts are getting into the system and could threaten the constraint(s).

After consideration of the dependent variables, the global impact of the measurements involved in the above effectiveness rating would be an immediate 15% decline in sales due to lost capacity from resource A $(.90 \times .95 - 1)$ and a 17% loss of sales due to scrap from resources A, B and C, for a total of 32% loss of sales that can never be recaptured. Inventory and Operating Expense would rise because of a lack of protective capacity at resource B. A good strategy would be to fix the scrap at resource C, then increase the availability of resource A. Obviously the need exists to fix both.

As with other conditions that interfere with the proper implementation of the five step improvement process, whenever a machine's status poses such a threat, corrective action should receive higher priority than usual. Additional insight into methods of focusing on those resources required in subordination through buffer management will be discussed later.

QUALITY FOCUS IN THE FACTORY

Where and how to focus a quality program in the factory is a matter of managing priorities, which are often determined by the availability of time and money. If the necessary condition of good quality, as perceived by the customer, is not being fulfilled, then priority must be given to solving this problem. The Quality Function Deployment (QFD) program defines the relationship between the individual functional departments and the customer. The objective is to ensure that those product attributes that are deemed unsatisfactory from the customer's perspective are fixed by the appropriate organization (see Chapter 9, "TOC and Process/Product Design").

The next order of business is to determine what portion of the factory is the most vulnerable with respect to Throughput, Inventory and Operating Expense, and to strengthen and protect those areas. Obviously, if the most heavily loaded resource has a quality problem, it must be dealt with immediately. This resource controls the Throughput generated for a large portion of the company, so solving a quality problem here can immediately improve profitability.

However, several questions must be answered before knowing where and how to focus activities to improve and protect profitability. For example,

- What is the impact of quality problems that occur at resources that are loaded to near or secondary constraint levels?
- What is the impact of constraints that process material that is already substandard and destined to be scrapped owing to operations at earlier resources?

Problems at Resources that Feed the Constraint

In Figure 3.9, resources A, B and C are connected by operations that feed linearly from left to right. Resources A and C have been loaded to 50% and have scrap rates of 20% and 15% respectively. Resource B, which has been identified as the constraint, is loaded to 100% and has a scrap rate of 5%.

Whenever A scraps a part, there is a loss of the raw material involved. Additional material must be started at the gating operation to replace what has been lost. The scrapping does not incur extra personnel cost, so the loss is the cost

```
                                          Scrap Rates
           A    B    C
*----*----*----*----*----*              A  =  20%
          50%  100% 50%                 B  =   5%
                                        C  =  15%
```

Figure 3.9 Problems at resources feeding the constraint.

of raw material alone. However, a high scrap rate indicates that a severe problem exists in a resource that directly feeds the constraint. If scrap material absorbs processing time on the constraint, Throughput will go down. While it may not be a high a priority to solve the scrap problem at resource A, it should be a very high priority to prevent, in any way possible, scrap material from being processed on resource B. If a method cannot be devised to screen out defective material, then fixing resource A would command a high priority.

Dealing with Near-Constraint Resources

Figure 3.10 is similar to Figure 3.9 except that instead of scrap, rework is the issue. The load on resources A and C has also been changed as a result of the additional labor required to rework the parts. Resource A's load has been increased to 90% resulting in a high probability that A will be unable to deliver to B to meet B's schedule.

To ensure that products that are to be processed at resource A will arrive at B on time, more processing time must be found for resource A. Under this condition, material is released earlier to ensure that parts can go from the gating operation to the constraint on time, resulting in an increase in Inventory, or else overtime must be expended, resulting in an increase in Operating Expense. So, fixing the rework at near-constraint resources will result in a decrease in either

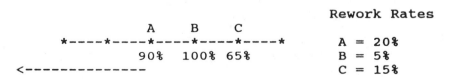

```
                                       Rework Rates
            A    B    C
      *----*----*----*----*----*        A  =  20%
           90%  100% 65%                B  =   5%
<---------------                        C  =  15%
```

Figure 3.10 Near-constraint resources.

Inventory or Operating Expense, or both, and an increase in the protection of the constraining resource (B).

Secondary Constraints

Figure 3.10 shows that if in trying to release material earlier at near-constraint resources it is found that time zero (today) were to prevent the release of material earlier (yesterday), then the rework would be responsible for creating the secondary constraint and a method would have to be devised for increasing available production time at resource A so that B could be protected. Secondary constraints also command a high priority.

Problems after the Constraint

Resource C, while it has a low utilization rate, processes material that has already been through the constraint. As seen earlier, whenever a part is scrapped here, there is a loss of raw material since additional material must be started as replacements, and there is a loss of constraint time. The time from the constraint used to create the part does not result in the generation of Throughput. In this case, fixing the scrap on resource C would also command a high priority.

But, what is the impact of rework that occurs after the constraint? In the following figure, resource B has a rework problem that has driven the load to near constraint levels at 90%. When this occurs, additional capacity must be found to ensure that orders will be processed on time. Failing that, sales orders will be pushed out and the product will arrive late to the customer.

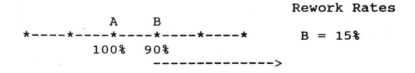

Fixing resource B's problem would result in less inventory between A and the end of the process and less Operating Expense used to support the overtime required to ship on time.

The Aggregation of Resource Capacity

Seldom will any manufacturing facility have only one resource. And so, when considering what must be fixed, it is important to consider the aggregation or total resource capability which ultimately determines the priorities. In other words, just because a resource is loaded to 85% does not automatically mean it should get attention. The following illustration suggests why a new perspective is necessary.

```
     A    B    C    D              Rework Rates
     *----*----*----*
     85%  30%  50% 100%            A = 15%
```

Resources A, B, C and D represent a chain of resources in a production process that feeds from resource A through to resource D. Resource D is the constraint, while resource A, because of a rework problem, is loaded to the near-constraint level (85%) and may have problems associated with an insufficient amount of capacity. But the question in this process is, will resource D be threatened with a lack of material arriving from A? Since the objective here is to ensure that D is used continuously in the production of Throughput, this question must be answered. Obviously, resource B will have problems in that it may not have enough available capacity but should money be spent to fix it? By looking at resource A from a local perspective, it appears that it needs to be fixed. But what is the *global* impact? Since the objective is to arrive at D on time, the impact of available capacity at all resources in front of D must be considered—not just A. So the question must be rephrased to ask what the impact is of available capacity on resources A, B and C in being able to deliver to D on time. With this in mind, the excess capacity available at B and C must be considered in making it possible to overcome any problem that may occur at A. This is the subject of buffer management and is discussed in Chapter 6, on improving Just-in-Time.

Conclusion

The most vulnerable portions of the factory with respect to quality and from a global perspective include:

- The primary and secondary constraint(s)
- The string of resources leading from the constraint(s) to the end of the process
- The material being fed to the constraint or constraints
- Near-constraint resources whose problems cannot be overcome by capacity at other resources.

Note that additional methods for increasing capacity at constrained resources include design/redesign of products and processes, statistical process control and design of experiments, all of which are discussed in later chapters.

• • •

If the constraint to making more money exists in an area other than production, no amount of reducing setup time or decreasing machine downtime will increase Throughput, and the chances are very great that a decrease in Operating Expense will not occur.

• • •

THE IMPACT OF PRODUCT FLOW ON THE SCHEDULING PROCESS

The relative positioning of constraints and non-constraints has a large impact on the way the organization performs, the problems that will surface, and the decisions that must be made regarding the TQM/SPC program. How a plant should be scheduled and the actions of workers and managers are directly affected by the way in which product flows through the factory. Certain situations with respect to the location of constraint and non-constraint resources will cause workers and managers to react in a predictable manner that will have a negative effect on profitability. Correcting these problems is imperative and is a primary tenet of the Drum-Buffer-Rope process discussed in Chapter 6, "Improving on Just-in-Time." The following is a synopsis of the situations that can occur, and the resulting impact.

Constraints feeding constraints

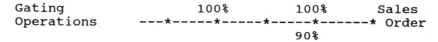

```
Gating                        100%              100%          Sales
Operations         ---*-----*-----*-----*------* Order
                                         90%
```

Whenever constraints of equal capacity occur in sequence, the first resource will be the primary constraint because of its inability to adequately feed the second. The maximum load will rarely be at maximum at the secondary constraint. However, at the secondary constraint, the amount of protection required to ensure that the sales order due dates will be met is greater and the predictiveness of sales order delivery is lower. Obviously, improvement programs should focus on maximizing the amount of Throughput that can be gained from the primary constraint and on increasing the amount of protective capacity available from the other. Fortunately, this situation of near-equal constraints does not exist very often. Because of the problems it creates, companies usually elevate one or the other.

Constraints feeding non-constraints

```
Gating                    100%     50%          Sales
Operations       ---*------*------*------* Order
```

When the constraint feeds a non-constraint operation, the non-constraint may be idle for long periods of time. The correct response in this case is to maximize the efficiency and utilization of the constraint, while subordinating those requirements of the non-constraint to whatever the constraint needs. However, because of faulty measuring systems, the general tendency is to attempt to work the non-constraint as well at high efficiency and utilization. Since obtaining benefits this way is impossible, managers will tend to reshape the environment by moving the constraint to attempt to increase Throughput. Constraints should be manipulated based on the desired impact on Throughput, Inventory and Operating Expense, and not to increase the efficiencies and utilization of non-constraint resources.

Non-Constraints feeding constraints

```
Gating                     50%    100%         Sales
Operations       *------*------*------* Order
```

As long as traditional measurements are used in this situation, the non-constraint resource will tend to be overutilized. Since there is no restriction to incoming material, it can keep producing. Unfortunately, overall output is controlled by the constraint. Inventories build and Operating Expense increases as a result. As inventories build, additional bottlenecks are created, requiring overtime.

Non-constraints feeding non-constraints

```
Gating                     50%     50%         Sales
Operations       ---*------*------*------* Order
```

Whenever a non-constraint feeds a non-constraint, the tendency will be to maximize the efficiency and utilization of each, resulting in an increase in Inventory and Operating Expense and decreasing Throughput. Non-constraints should be utilized only when necessary to feed constraints or in the creation of Throughput.

Non-constraints and constraints feeding assembly operations (Convergent)

```
                              A       C
                            100%    Asmb           Sales
Gating                   ---*-----*------*------* Order
Operations                               |
                         ---*-----*------*
                            50%
                             B
```

In convergent operations, regardless of whether a constraint is involved, the tendency is to maximize the utilization of all resources by using large lot sizes, The result is that material will move through the facility in waves. Resources will be overloaded one minute and empty of work the next.

In cases where one leg of the convergence is restricted, as in the following illustration, the overutilization of the non-constraint will force work in process to increase in the assembly operation while underprotection of the non-constraint leg will leave material arriving from the constraint leg sitting in assembly without matching parts. Resource A is the constraint with a 100% load, while resource B is the non-constraint with a 50% load. Overutilizing resource B will cause material to collect in resource C, which cannot go further without parts from A. However, parts arriving from resource A should not be kept waiting at resource C for matching parts from resource B.

Non-constraints and constraints feeding more than one operation (Divergent).

```
                         A       B
Gating                ---*-----*------*------* Downstream
Operations                     |             Operations
                          ------*------*
                               C
```

Whenever there is a divergence in the delivery of material, there is an opportunity for misallocation. Material that should be used to fill one order can end up used on another or returned to stock. Material flowing from resource A diverges and is sent to resources B and C. If B is overutilized, it will take more material from resource A than is needed, and C will not be able to supply parts to downstream operations. In companies where there is a large amount of diverging operations, misallocation can be a major problem. To prevent a misallocation of material resources, B and C should be provided with schedules to tell these operations to produce a certain amount of material of a specific kind and then stop.

Analyzing Product Flow Types

Once an understanding of the impact of specific product flow has been achieved, determining what the primary flow type is for a specific facility will help tremendously in being able to predict the types of problems that are prevalent and then in providing a solution.

Plants are classified as A-, V- or T. The letters refer to the shape of the schematic depiction of product flow within the plant.

A-Plants

The A-Plant is characterized by a large number of converging operations starting with a wide variety of raw material items being assembled in succeeding levels to create a smaller number of end items. Components are usually unique to the end item and the technology used in assembly operations tends to be highly flexible, general-purpose equipment. Under traditional management practices, the tendency is to misallocate resource time in an attempt to maximize efficiency and utilization. Large batches are used to keep these figures high resulting in a poor component mix and a constant shortage of the right parts at assembly operations. These large batches move in waves throughout the plant, causing temporary bottlenecks to wander from resource to resource. Machines may be underutilized one minute and overutilized the next. Since material is constantly out of balance, overtime is used to "catch up" so that shipments can be made on time. The resulting impact on the quality system is disastrous. Large batches will result in an increase in Inventory and a reduction of visibility. Quality will tend to decrease. Because of the poor availability of materials and constant expediting, there is much pressure exerted to pass marginal material as acceptable to meet scheduled due dates.

In order to deal with these problems, process batch sizes must be changed so that they maximize utilization at the constraint(s), while transfer batches should be made as small as possible. The sequencing and timing of individual orders across all operations should be synchronized with the schedule created for the constraint operation(s). Operators and foreman are encouraged to activate resources to produce only what is required. Buffers designed to protect the Throughput of the system from those things that can go wrong are used for shipping, constraint and assembly operations (see Chapter 6).

The typical A-Plant is associated with a manufacturer of complex make-to-order products. Figure 3.11 is indicative of the A-Plant product flow structure for one product.

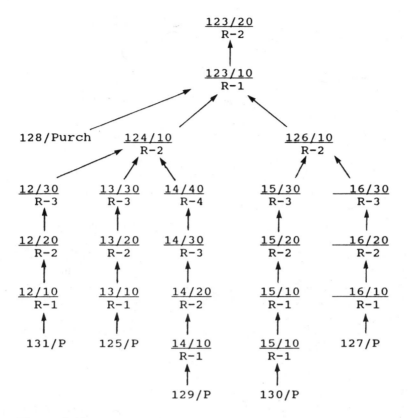

Figure 3.11 A-plant product flow.

V-Plants

V-Plants are characterized by constantly diverging operations, with a small number of raw material items being converted into a large number of end items, using highly specialized and expensive equipment. Since each diverging point in the process is an opportunity to misallocate material, the V-Plant, unlike the A-Plant, is dominated by this problem. Under the traditional approach, expensive equipment must be utilized constantly to absorb overhead and to insure that adequate value is received. Setups are extensive, so batch sizes will be large. Unfortunately, this results in material being taken from diverging operations in quantities larger than required. One leg of the divergence will be unable to perform because another leg has received material that should have gone to it. Misallocation of material prior to the constraint results in a misutilization of

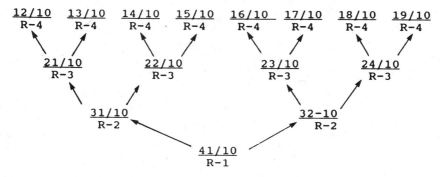

Figure 3.12 V-Plant product flow.

constraint time. Material processed at the constraint that is not dedicated to creating Throughput results in a decline in ROI. Misallocation of material after the constraint will result in an increase in finished goods or products for which there is no demand. Customer service levels will be poor. To offset for constantly being out of stock, finished goods inventories will be raised even higher through an attempt to forecast. Like the A-Plant, large batches increase Inventory and reduce visibility, and quality suffers. The solution, which is much the same as that for the A-Plant, is to synchronize product flows with the systems constraints and customer demand. Lot sizes should be matched to requirements for creating Throughput, while minimizing Inventory and Operating Expense.

The forged products manufacturer is typically associated with V-Plants, as represented by the product flow shown in Figure 3.12.

T-Plants

T-Plants are characterized by a relatively low number of common raw material and component parts optioned into a large number of end items. To support a requirement for meeting short lead time demand a two level master schedule is normally used, in which common components are scheduled and stored just prior to final assembly via forecast and then assembled to order based on specific customer configuration. The T-Plant distinguishes itself from the A-Plant in that the A-Plant is dominated by the convergence interaction, whereas the T-Plant is dominated by the divergence that occurs just prior to final assembly. Prior to this stage, there are no converging or diverging operations. Raw material is processed without being assembled or converted into more than one part, so the quantities of raw material and subassembly components quantities will be the same.

Since diverging operations give opportunity for the misallocation of material, inventories at final assembly will not match customer demand. Customer service levels will be low. Under traditional management strategies, better equipment utilization means larger lot sizes, resulting in the same wave effect seen in A-Plants, exacerbating the out-of-balance conditions in inventories while extending lead times and reducing visibility. Quality suffers and attempts at "modernizing" the plant and bringing in "new," "more efficient" and "cost-effective" equipment may result in less flexibility and an even bigger desire to maximize equipment utilization, making the problem even worse.

T-Plants are best run as two separate facilities: one for make-to-order and the other for make-to-stock parts, as in traditional two-level master scheduling. Product flow and demand should be synchronized to the constraint and market demands as in the other plant types, and Inventory or "stock" buffers should be placed at final assembly and time buffers at assembly, shipping and constraint operations.

The consumer products manufacturer is an example of a T-Plant operation, in which common components are created using similar processes and then assembled based on the desired customer requirements. Figure 3.13 illustrates the T-Plant flow structure.

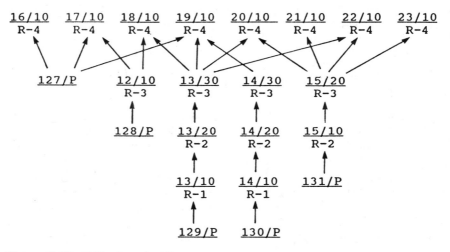

Figure 3.13 T-Plant product flow.

Combination Plants

Companies do not always fall neatly into the A-, V- or T-Plant categories. They may have characteristics of each, in various combinations. As an example, a forged products facility may be used to feed an assembly plant. This would be an example of a V-Plant feeding an A-Plant. A customized industrial computer manufacturer may purchase raw components for assembly to stock, and then final assembly to order. This would be an example of an A-Plant feeding a T-Plant.

The product flow diagram gives a detailed record of the relationship of all part/operations and their associated resources to identify characteristics that are inherent in A-, V- and T-Plants and enable one to understand what instructions are necessary for people who work in specific areas of the production process. As an example, workers in operations just after divergent stations can be made aware that a major problem is misallocation and that only material required to fulfill the schedule should be taken.

THE INTRODUCTION OF THE CONSTRAINT-BASED INFORMATION SYSTEM

Determining the relationships among resources and their impact on each other is the role of the new constraint-based information systems designed for the dependent-variable environment. Following this approach allows one to address some interesting questions:

- What is the impact if a part is scheduled at a resource that has minimum protective capacity and that is between the constraint and the delivery of the sales order?
- What is the impact if a part is scheduled at a resource that has minimum protective capacity and that is between the constraint and the release of raw material?

The TOC-compatible information system will be discussed further in Chapter 13.

• • •

While understanding the physical environment is critical to supporting a process of continuous profit improvement, it is extremely important to remember that most constraints are not physical. They are created by invalid policies that serve to inhibit the ability to maximize resource utilization in the creation or protection of Throughput.

• • •

STUDY QUESTIONS

1. What is the impact of statistical fluctuations on the dependent-variable environment?
2. What is a product flow diagram, and how is it distinguished from a process flow diagram?
3. What is a part/operation or station, how is it used and how is it constructed?
4. Define the single-minute exchange of die (SMED) process and give an example of using cost as a basis for focusing improvements.
5. What is the distinction between a constraint and a bottleneck?
6. What is inertia and why is it important?
7. What is total productive maintenance (TPM) and what is the impact of the effectiveness rating.
8. What is the impact of scrapping a part after it has been processed at the constraint?
9. What is the impact of scrapping a part before it reaches the constraint?
10. Define the following: A-plant, T-plant, V-plant combination plant.
11. What negative effects would be found as a result of maximizing resource utilization in the A-, T- and V-plant environments?
12. What is meant by the concept of the aggregation of resource capacity?
13. What is meant by the concept of the aggregation of Murphy?
14. What is the primary cause of constraints?

4

Analyzing Policy Constraints:
The TOC Thinking Process

This chapter introduces the concept of policy analysis and provides a method to develop simple solutions.

OBJECTIVES

- To understand the basic concept of policy analysis
- To present a process for analyzing poor policies
- To begin to understand how to overcome policy problems associated with the five-step improvement process discussed in Chapter 2 so that constraints can be exploited and other resources can be subordinated to them, and so that constraints can be elevated when desired
- To begin to understand how to create simple solutions

THE CONCEPT OF POLICY ANALYSIS

Perhaps the most limiting factor to a company's ability to make money is its internal policies. Poor policies inhibit the maximization of Throughput from existing constraints and create false assumptions about suboptimization. As an example, a policy of maximizing the utilization of every resource, instead of maximizing profitability, would ultimately cause bankruptcy by increasing inventories and interfering with a physical resource's ability to create Throughput. Economic order quantity (EOQ) assumes that a balance of suboptimization must be struck between the cost of setup and the cost of carrying inventory. But setup costs are not constant. The cost of setup at one work center may impact Throughput while the cost at another may not exist at all.

. . .

Because there will always be a constraint to a company's making more money, the probability is 100% that at some point the company will arrive at a constraint that they have no experience with breaking.

. . .

To identify and solve these problems takes more than an analysis of the physical environment. A diagnostic capability, while useful in gaining additional insight, cannot identify non-physical limitations. What is needed is the capability to analyze effects that occur in the environment so that core causes can be discovered and simple solutions developed.

CREATING THE ROAD MAP*

The objective of the TOC thinking process is to define those actions necessary to improve a company, given its current situation, and to guide each step to a sometimes not so obvious conclusion. It defines:

- *What* to change
- What to change *to*
- *How* to accomplish the change

Defined in *It's Not Luck* (Goldratt, 1995) and included in the TOC thinking process are five tools, including:

- *Currently reality tree (CRT)*—used to find the core cause or causes from undesirable effects (UDEs)
- *evaporating cloud (EC)* assumption model—used to model the assumptions that block the creation of a breakthrough solution
- *Future reality tree (FRT)*—used to model the changes created after defining breakthrough changes from the evaporating cloud
- *Prerequisite tree (PT)*—used to uncover and solve intermediate obstacles to achieving the goal
- *Transition tree (TT)*—used to define those actions necessary to achieve the goal

The TOC thinking process is used to guide the implementation of the theory of constraints and to aid in creating breakthrough solutions. It can also be of tremendous benefit to the implementation of the TOC-based system by ensuring that any undesirable effects that manifest themselves during the implementation process are eliminated. [An excellent reference for the TOC thinking process is

The Theory of Constraints: A Systems Approach to Copntinuous Improvement (Goldratt, 1995).]

THE CURRENT REALITY TREE

The first step in the execution of the TOC thinking process is to list the undesirable effects (UDEs) and then create the current reality tree (CRT). The CRT is a diagram built on the cause-and-effect relationships between the undesirable effects and their immediate causes. The objective is to find the core cause. Once found and eliminated, all undesirable effects should disappear.

As an example, the following are undesirable effects that may occur during implementation of a TOC-based information system and can be used to build a CRT. If the core cause can be found and a solution implemented, then these effects should no longer exist. If they continue, then the core cause has not been eliminated.

UDE-1   Resource R-3 often receives mismatched parts from R-4.

UDE-2   Sales orders are frequently shipped late for product A.

UDE-3   Resource R-2 often has holes in zone 1 of the buffer caused by late orders for part/operation A/10.

UDE-4   Part/operations B/30 and D/30 are often expedited from R-4

UDE-5   There are holes in zone 1 of the shipping buffer for product A

UDE-6   R-4 is consistently working overtime to supply parts D/30 and B/30

Notice that each UDE has been numbered. This is done so that each UDE can be easily identified in a CRT. Figure 4.1 is the product flow diagram showing the

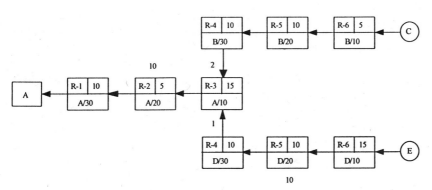

Figure 4.1 The product flow diagram for product A.

previously defined UDEs. Note that the product flow diagram is not part of the thinking process. It has been supplied as a reference point during the discussion of the five tools.

In Figure 4.2, UDE-2 is the first undesirable effect to be considered and appears at the head of the arrow. In other words, it is an identifiable effect and can be related directly to the text in block 1. Block 1 and all other numbered blocks are known as "entities." To read the diagram, the block at the tail of the arrow is read first. The tail of the arrow is identified with the word "if" while the head of the arrow is identified with the word "then." So the relationship between UDE-2 and entity-1 should read:

> **if:** (entity 1) product A is frequently late arriving from production
>
> **then:** (UDE-2) sales orders are frequently shipped late for product A

Entity-1 is also connected to UDE-5.

> **if:** (entity 1) product A is frequently late arriving from production
>
> **then:** (UDE-5) there are holes in the zone 1 of the shipping buffer for product A

The process of identifying relationships between UDEs and entities continues until the core cause is discovered.

UDE-3 is also connected to entity-1. However, UDE-3 is at the tail of the arrow. There is now a connection between UDEs 2, 3 and 5 through entity-1.

> **if:** (UDE-3) resource R-2 often has holes in zone 1 of the buffer caused by late orders for part/operation A/10
>
> **then:** (entity-1) product A is frequently arriving late from production

Notice that this statement cannot necessarily stand by itself. Just because there are holes in the buffer for product A does not mean that product A will be frequently late in arriving from production. There must be supporting causes. Entity-2 states that orders for part/operation A/10 cannot be expedited in time to meet the demand of the constraint buffer. Note that the constraint has been identified in the product flow diagram as R-2.

The arrows originating from UDE-3 and entity-2 are joined by an ellipse along with entity-3. In this case, the relationship is read:

> **if:** (UDE-3) resource R-2 often has holes in zone 1 of the buffer caused by late orders for part/operation A/10

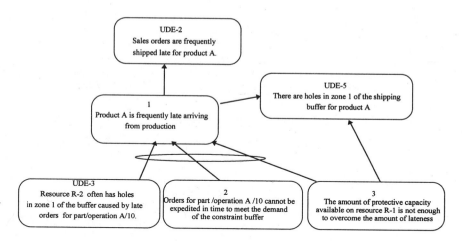

Figure 4.2 Current reality tree: entities-1–3, UDEs-2, -3, amd -5.

> **and if:** (entity-2) orders for part/operation A/10 cannot be expedited in time to meet the demand of the constraint buffer
>
> **and if:** (entity-3) the amount of protective capacity available at resource R-1 is not enough to overcome the amount of lateness
>
> **then:** (entity-1) Product A is frequently arriving late from production

In this case, there is enough evidence to point to a definite cause-and-effect relationship between entity-1, UDE-3, entity-2 and entity-3.

The CRT can provide tremendous insight into the cause-and-effect relationships of activities occurring throughout the plant. While it is developed from the undesirable effects and projects downward toward the core causes, the CRT can, and should, be validated by reading from the bottom up. Entity-14 in Figure 4.4 has been identified as the core cause for creating the late orders. When combined with entity-15, there is sufficient cause-and-effect relationship to justify entity-7.

> **if:** (entity-14) R-4 is combining lots to save setup
>
> **and if:** (entity-15) R-4 has not been identified as a secondary constraint and a schedule produced to subordinate R-4 to R-2
>
> **then:** (entity-7) R-4 is not producing B/30 and D/30 according to schedule

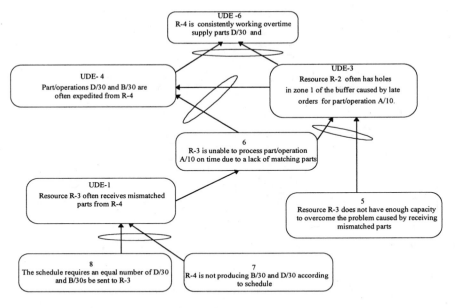

Figure 4.3 Current reality tree: entities-5–8, UDEs-1, -3, -4 and -6.

The issue now turns to understanding why setup savings are being performed for part/operations B/30 and D/30 at resource R-4. Before a successful plan can be adopted, a basic understanding of *why* is important. Short of asking the foreman, a basic analysis should be performed. (Asking the foreman why he is performing setup savings on a non-constraint resource may only solidify his insistence to continue.)

Figure 4.5 presents one reason that the foreman may be acting as he is. Beginning at the bottom it reads:

> **if:** (entity-18) people act based on how they are measured
> **and if:** (entity-19) R-4 is measured based on productivity
> **then:** (entity-17) the actions of people at R-4 will be to maximize the productivity at R-4

Continuing to the next level, the tree reads:

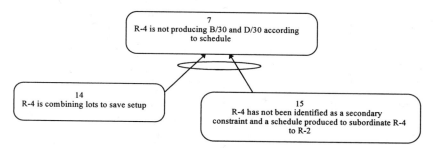

Figure 4.4 Current reality tree: entities-7, -14 and -15.

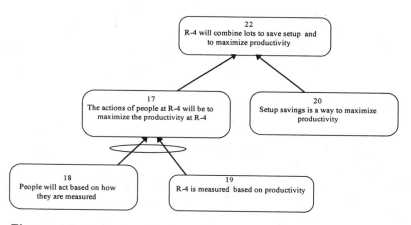

Figure 4.5 Current reality tree: entities-17–20 and -22.

if: (entity-17) the actions of people at R-4 are to maximize the productivity at R-4

and if: (entity-20) setup savings is a way to maximize productivity

then: (entity-22) R-4 will combine lots to save setup and maximize productivity

So the core cause for the setup savings occurring an a non-constraint resource may be how the foreman is being measured externally (a managerial constraint), or in how he measures himself (a behavioral constraint).

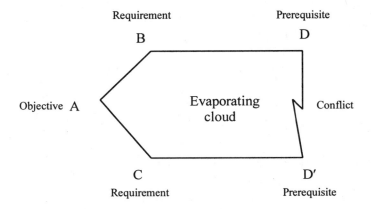

Figure 4.6 Evaporating cloud.

CREATING THE EVAPORATING CLOUD

As stated earlier, the evaporating cloud (EC) is used to model those assumptions that block the creation of a breakthrough solution (Figure 4.6) Notice that the objective of the model is placed on the left at position A. The objective may be to increase return on investment. Positions B and C represent those things that are required to attain the objective. Position B might read *increase productivity* and position C might read *exploit the constraint*. Positions D and D′ represent the conflict. There is an assumption that D and D′ cannot exist together.

The cloud reads as follows:

- In order to have A there must be B
- In order to have A there must be C
- A prerequisite to B is that there must be D
- A prerequisite to C is that there must be D′
- D and D′ cannot exist together

To solve a problem involving the evaporating cloud, the assumptions made between each of the entries in the diagram are examined for possible flaws. In Figure 4.7, in order to say that there is a relationship between A and B, certain assumptions must be made, such as the only way to have A is to first have B. If the assumptions can be proven false, then the problem evaporates.

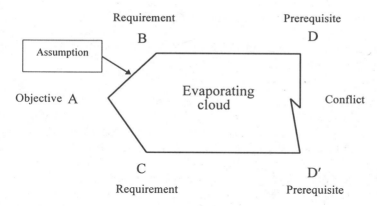

Figure 4.7 Evaporating cloud showing assumption.

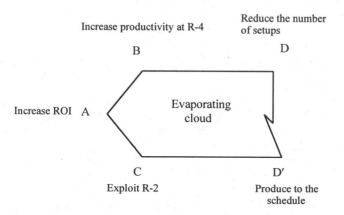

Figure 4.8 Evaporating cloud showing the foreman's dilemma.

THE FOREMAN'S DILEMMA: EVAPORATING THE CLOUD

Figure 4.8 represents the foreman's dilemma. In this case the objective is to increase return on investment.

The foreman's cloud reads as follows:

- A-B: in order to increase return on investment, productivity at R-4 must be increased
- A-C: in order to increase return on investment, R-2 must be exploited
- B-D: in order to increase productivity at R-4, the number of setups must be reduced
- C-D': in order to exploit R-2, R-4 must produce to the schedule
- D-D': producing to the schedule set for R-2 and saving setup are mutually exclusive

To solve the foreman's dilemma the first step is to list the assumptions being made between the entries. Most of the time there will be more than one assumption made. The assumptions are:

- A-B: that there is a direct relationship between the local measurement of productivity at R-4 and the global measurement of return on investment
- A-C: that by exploiting R-2, Throughput will increase, thereby causing ROI to increase
- B-D: that setup savings at R-4 can only have a positive impact
- C-D': that R-4 must be subordinated to the demand created by R-2
- D-D': that there is no way that R-4 can follow the schedule for R-2 and perform setup savings at the same time

A-B makes the assumption that there is a direct relationship between the increased productivity at resource R-4 and the increase in return on investment. This assumption is not necessarily true. Improving at a non-constraint resource from a global perspective will not cause Throughput to increase. B-D also has a problem. As will be seen later, saving setup at a non-constraint resource will often subvert the schedule of the constraint, causing ROI to decrease rather than increase. If setup savings is not performed in direct synchronization with the constraint's schedule, then there is a good chance that the constraint schedule will be invalidated by the setup process. This is a key issue in solving the foreman's dilemma.

FUTURE REALITY TREE

As defined earlier, the future reality tree (FRT) is used to model the changes created after defining the breakthrough solution from the evaporating cloud. In this case, a new measurement system is being provided to the foreman (Figures 4.9 to 4.11). The FRT is built starting with what is called an injection and is read in the same manner as the CRT. An injection is the change designed to solve the undesirable effects. In this case the injection is to measure the foreman's performance based on throughput dollar days (T$D). (For a discussion of throughput dollar days, see p. 119 in Chapter 6.) The FRT reads as follows.

> **if:** (injection-1) the foreman's performance is measured based on Throughput dollar days (T$D) accumulated
>
> **then:** (entity-26) pressure will increase on the foreman to deliver parts according to the schedule for R-2
>
> **if:** (entity-26) pressure will increase on the foreman to deliver parts according to the schedule for R-2
>
> **and if:** (entity-27) the schedule for R-2 is communicated to R-4 through the release schedule for R-6
>
> **then:** (entity-28) R-4 will begin to process orders as they are received (FIFO)

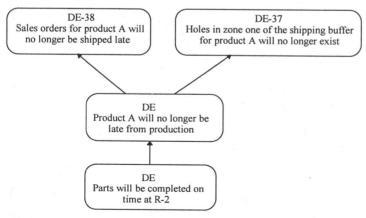

Figure 4.9 Future reality tree (entities 35–38).

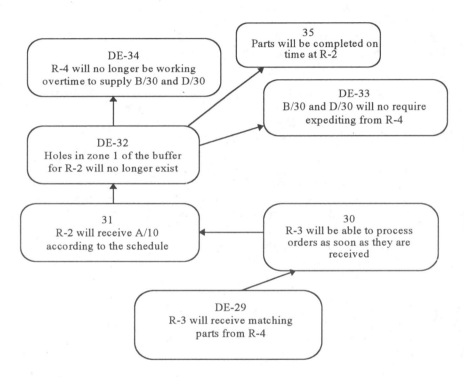

Figure 4.10 Future reality tree (entities 29–35).

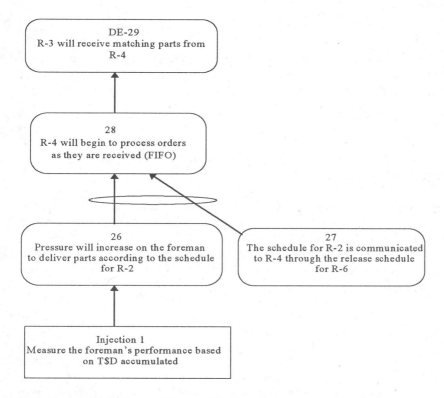

Figure 4.11   Future reality tree (injection 1 and entities 26–29).

Notice that entity-29 is a desirable effect (DE). This means that the entity has the exact opposite of the corresponding undesirable effect from the UDE list. In this case, one of the undesirable effects has been eliminated by a change in the way the foreman is measured.

> **if:** (entity-28) R-4 will begin to process orders as they are received (FIFO)
> **then:** (DE-29) R-3 will receive matching parts from R-4

Notice that in the remainder of the CRT, all of the undesirable effects have been eliminated, including UDE-2 (sales orders are frequently shipped late for product A) and replaced by desirable effects.

THE FOREMAN'S DILEMMA: TRIMMING THE NEGATIVE BRANCHES

Somehow the solution does not seem complete. If the measurement system was not the only issue preventing the foreman from following the schedule, then the devised solution may not replace all the undesirable effects with desirable effects as defined in the future reality tree. At this point, one may need to question the validity of the solution. Figure 4.12 serves to verbalize the reservation.

> **if:** (injection-1) the foreman's performance is measured based on T$D accumulated
> **and if:** (entity-27) the foreman firmly believes that the problem associated with the lateness of orders is due to a lack of adequate capacity at resource R-4
> **then:** (entity-25) the foreman will continue to perform setup savings for B/30 and D/30

> **if:** (entity-25) the foreman will continue to perform setup savings for B/30 and D/30
> **then:** (entity-26) the undesirable effects will not go away

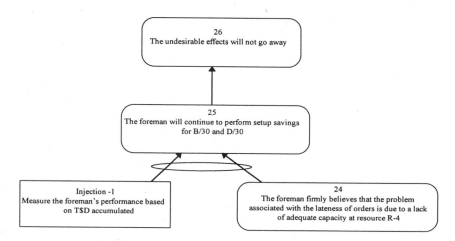

Figure 4.12 Identifying reservations.

REVISITING THE FOREMAN'S DILEMMA

From Figure 4.12 it is learned that there is a distinct possibility that the foreman firmly believes that the undesirable effects are caused by insufficient capacity at R-4. Merely placing a new measurement system in this environment may only serve to exacerbate the situation. Figure 4.13 defines another possible cause for the setup savings. It reads as follows:

if:	(entity-23) a constraint will sometimes identify itself by the amount of product being expedited or the amount of overtime required
and if:	(UDE-4) part/operation D/30 and B/30 are often expedited from R-4
and if:	(UDE-6) R-4 is consistently working overtime to supply parts D/30 and B/30
then:	(entity-19) R-4 is perceived to be a secondary constraint
if:	(entity-18) people act based on their perceptions
and if:	(entity-19) R-4 is perceived to be a secondary constraint

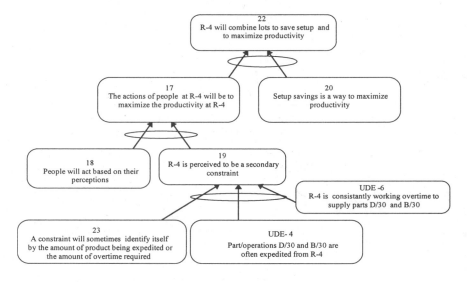

Figure 4.13   Validating the reservation in Figure 4.12.

     then: (entity-17) the actions of people at R-4 will be to maximize the productivity at R-4

       if: (entity-17) the actions of people at R-4 will be to maximize the productivity at R-4

and if: (entity-20) setup savings is a way to maximize productivity

  then   (entity-22) R-4 will combine lots to save setup and to maximize productivity

EXPANDING THE FUTURE REALITY TREE

To solve the new problem, the FRT must be expanded to include a new entity. The bottom of the FRT now reads:

       if: (injection-1) the foreman's performance is measured based on T$D accumulated

    then: (entity-26) pressure will increase on the foreman to deliver parts according to the schedule for R-2

    if: (entity-26) pressure will increase on the foreman to deliver parts according to the schedule for R-2

    and if: (entity-26) the schedule for R-2 is communicated to R-4 through the release schedule for R-6

    and if: (entity-40) the foreman believes that R-4 is not a secondary constraint

    then: (entity-28) R-4 will begin to process orders as they are received (FIFO)

This is shown in Figure 4.14.

In order for the problem to be solved, the foreman must believe that R-4 is not a secondary constraint. A new injection must be introduced that will cause the

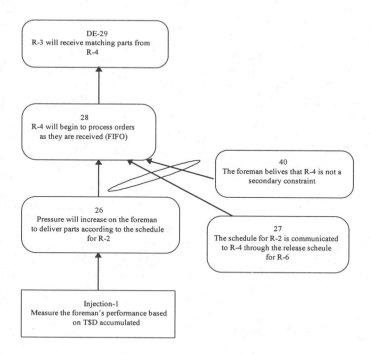

Figure 4.14   Recognizing the new entity (29).

foreman to change his belief. Assuming that the implementation of the system has included education and training for the foreman, he should already be aware of the impact of setup savings on a non-constràint resource. The solution may be in providing a method for the foreman to simulate his current environment to enable him to either validate or disprove his belief. This issue serves to underline the need for a permanent education program that will help people invent their own solutions.

SIMULATING THE FOREMAN'S PROBLEM USING THE INFORMATION SYSTEM

Figure 4.15 shows why the parts arriving at R-3 are mismatched. If setup savings is performed on R-4, some of the orders will be pushed outward in time, while others will be pulled in. The result is that R-4, which had been synchronized to the constraint schedule of R-2 by the release schedule for R-6, will no longer be synchronized. According to the schedule in Figure 4.15, R-3 will receive large batches of B/30 and then large batches of D/30. The foreman begins working overtime to keep up with material expedited to R-3. His conclusion is that he does not have enough available capacity to remedy the situation. This is a normal conclusion and is made every day in the traditional MRP facility.

Figure 4.15 reveals exactly the kind of information that the foreman needs to be induced to change his activity. Fortunately the TOC-compatible information system defined in Chapter 13 is capable of simulating the interface requirements between R-2 and R-4. If the constraint schedule has been created for R-2 and it becomes necessary to determine the impact on R-4, R-4 can be declared a secondary constraint. Orders are then placed on the time line in synchronization to R-2. If any conflicts exist, they will surface as two parts/operations trying to occupy the same space at the same time on the time line for R-4 (Figure 4.16).

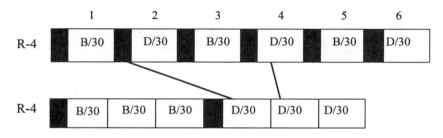

Figure 4.15 Simulating the foreman's problem.

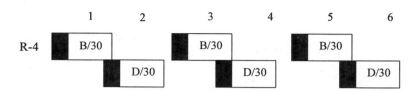

Figure 4.16 Placing orders on the time line for R-4.

Notice that orders 1 and 2 are trying to occupy the same space on the time line. However, this alone does not mean that there is a problem. To understand the implications of this condition, a schedule needs to be created for R-4 (Figure 4.17).

If the system places orders so that there are no gaps between orders (no additional time), this is an indication that R-4's time is limited. If there are no late orders and setup savings is not indicated on the system, then chances are good that setup savings is not required on the shop floor either. Gaps between orders on the schedule for R-4 are an indication that there is plenty of additional time and that the foreman's assumption that R-4 should be declared a secondary constraint and setup savings performed is false (Figure 4.18).

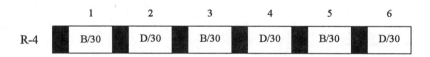

Figure 4.17 Generating the schedule.

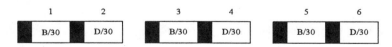

Figure 4.18 Adding the new injection.

Notice that there is a gap between the start of the planning horizon and the start of order 1. Notice also that there are gaps between orders 2 and 3 as well as orders 4 and 5.

ADDING THE NEW INJECTIONS

There are actually two new injections that are necessary to induce the foreman to stop setup savings at R-4 (Figure 4.19).

if: (injection-2) the foreman is provided with a method to simulate his environment

and if: (entity-43) the simulation indicates that R-4 is not the constraint

then: (entity-40) the foreman believes that R-4 is not a secondary constraint

Figure 4.20 contains injection-3.

if: (injection-3) the foreman is provided with training in the DBR process

then: (entity-41) the foreman will understand the impact of setup savings on a non-constraint resource

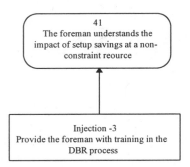

Figure 4.20   Adding the new injection.

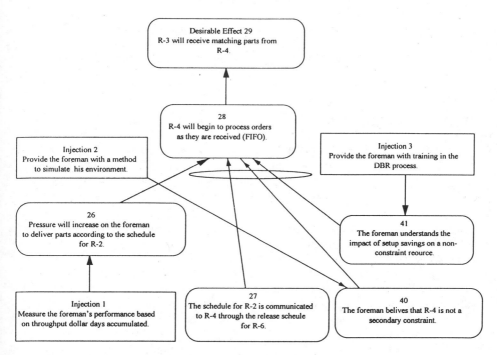

Figure 4.21 Expanding the future reality tree.

In Figure 4.21, the bottom of the FRT now reads:

if:	(entity-26) pressure will increase on the foreman to deliver parts according to schedule
and if:	(entity-27) the schedule for R-2 is communicated to R-4 through the release schedule for R-6
and if:	(entity-40) the foreman believes that R-4 is not the secondary constraint
and if:	(entity-41) the foreman understands the impact of setup savings on a non-constraint resource
then:	(entity-28) R-4 will begin to process orders as they are received

THE PREREQUISITE TREE

As stated earlier, the objective of the prerequisite tree (PT) is to uncover and solve intermediate obstacles to achieving the goal. The injections from the FRT are used to guide the process. As each injection is presented, obstacles to its realization surface. The intermediate objective in attaining the injection is to overcome the obstacle.

The injection is usually placed at the top of the diagram in a rectangle, the obstacle is placed to the left and is contained in a hexagon, while the intermediate objective is contained in a rounded rectangle.

Figure 4.22 involves injection-1 (measure of the foreman's performance based on T$D accumulated). Notice that one of the obstacles (obstacle-2) to attaining injection-1 is that the foreman may not understand the benefit of the T$D measurement process. This could easily derail the benefits of using the new measurement system. The other obstacle (obstacle-4) to attaining injection one is that the tools necessary to collect and manipulate the T$D data for the new measurement system are not available. To overcome the obstacle, the intermediate objective is used. The intermediate objective is the exact opposite of the obstacle. The intermediate objective to overcome obstacle-2 is that the foreman understands the measurement system. The opposite of obstacle-4 is that the tools are available. It now becomes clear that in order to attain injection-1, action must be taken to make sure that the foreman understands the measurement system and that tools are made available to collect data.

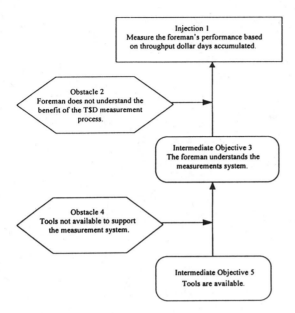

Figure 4.22   The prerequisite tree (injection-1).

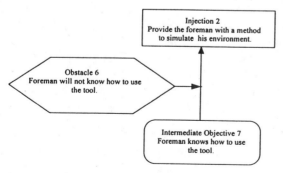

Figure 4.23   The prerequisite tree (injection-2).

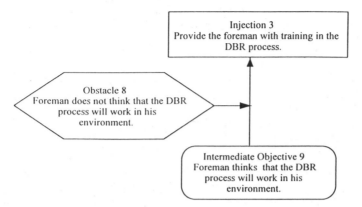

Figure 4.24   The prerequisite tree (injection-3).

In Figure 4.23 the obstacle to attaining injection-2 is that the foreman will not know how to use the tool. The intermediate objective is that the foreman does know how to use the simulation tool.

In Figure 4.24 the obstacle to attaining injection three is that the foreman does not think that the DBR process will work in his environment. The intermediate objective is that the foreman thinks the DBR process will work in his environment.

TRANSITION TREE

As defined earlier, the transition tree is used to define those actions necessary to achieve the goal. Like the prerequisite tree, the transition tree begins with the injection (Figure 4.25). The intermediate objectives are listed to the left in the rounded rectangular boxes and the actions designed to attain the intermediate objective are placed in rectangles. The arrows designate which actions apply to the intermediate objective. In order to attain intermediate objective-5, action-1 must take place. In order to have intermediate objective, three action two must take place.

The two actions required to attain the intermediate objectives as well as injection-1 include training the foreman so that he understands the measurement system and creating the tools to collect and manipulate the T$D data from the shop floor.

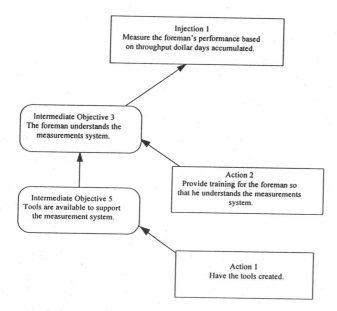

Figure 4.25 The transition tree (injection-1).

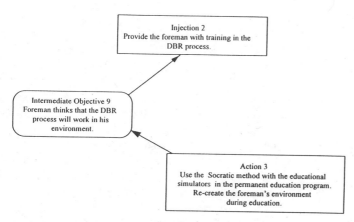

Figure 4.26 The transition tree (injection-2).

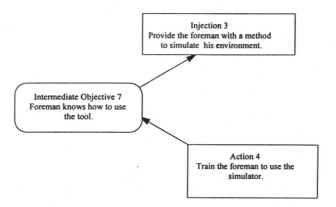

Figure 4.27   The transition tree (injection-3).

In Figure 4.26 the action required to ensure that the foreman believes that the DBR process will work in his environment is to use the Socratic method with the educational simulators in the permanent education program (action-3) and re-create the foreman's environment during education.

The action required to ensure that the foreman knows how to use the simulation tool is train the foreman to use the simulator (action-4) (Figure 4.27).

The TOC thinking process (TP) has been used successfully in companies around the world to create solutions to very complex problems. Using TP to aid in the implementation of the manufacturing information system is a logical extension.

USING THE INFORMATION SYSTEM TO DEVELOP THE ROAD MAP

Since the TOC-compatible information system is an excellent representation of reality and is devoid of policy constraints, it can be an excellent tool for helping to develop and execute the road map. As an example, the physical environment and its attributes will have a significant impact on solutions provided by the road map. The information system will not only inform the user as to what the impact of the physical attributes might be but can be used to test specific changes suggested in the creation of a breakthrough solution and will also aid in its execution.

There are several ways in which the information system can be used in support of the generation of the road map.

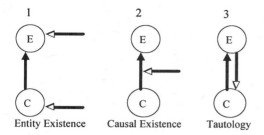

Figure 4.28   Categories of legitimate resevation.

- *Current reality tree*: Does excess capacity exist? If so, where?
- *Breakthrough solution*: Testing assumptions
- *Future reality tree*: If a specific action is taken then . . .
- *Prerequisite tree*: If a specific goal is identified what intermediate obstacles will block attaining the goal? (As an example, analyze conflicts between resources)

CATEGORIES OF LEGITIMATE RESERVATION (CLR)

Categories of legitimate reservation (CLR) refer to those problems related to the validity of certain connections between cause-and-effect relationships. There may be questions of whether or not an entity in fact exists (entity existence). Other questions may surface about whether an effect was created as a result of a given cause (causal existence). When reviewing a set of diagrams, these reservations can be easily identified (Figure 4.28).

Diagram 1 in Figure 4.28 shows a question of entity existence for both the cause and the effect. Diagram 2 shows a question of causal existence. The dark arrow signifies the relationship while the white arrow indicates the reservation. Diagram 3 represents tautology. In a tautology the cause is just a restatement of the effect. For example, the "customer must obtain a lower price" in order that he can preserve his profit margin. The prerequisite shown below suggests that the prerequisite is "must get lower price." The prerequisite is an obvious restatement

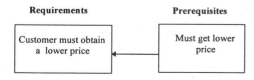

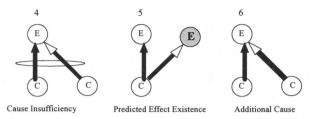

Cause Insufficiency   Predicted Effect Existence   Additional Cause

Figure 4.29 Categories of legitimate reservation.

of the requirement. Therefore the relationship suffers from tautology. The cause-and-effect relationship has not been thoroughly examined.

CLR exist to help perfect the diagrams explained earlier. Once a reservation has been raised, it may be answered by acknowledging the lack of relationship, or it may need an additional supporting cause (Figure 4.29).

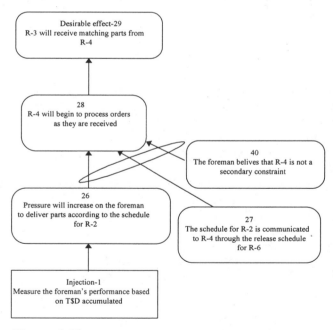

Figure 4.30 Cause insufficiency.

Cause Insufficiency

Cause insufficiency indicates that the existence of an effect cannot be explained by just one cause. In Figure 4.30 entity-28 (R-4 will begin to process orders as they are received) cannot be explained by just entity-26 (pressure will increase on the foreman to deliver parts according to the schedule for R-2). In order to fully explain entity-28, entities-27 and -40 must also be present.

Predicted Effect Existence

Predicted effect existence can be very helpful in establishing relationships in which a specific cause is speculated but is not directly attributable to the effect. The objective is to determine whether a specific cause exists. If a direct link cannot be made between two entities and another effect can be readily observed and is known to have a direct link to the cause, this effect is used to support the existence of the original cause. This term grew from the effect-cause-effect concept (early TOC concept) in which an effect was viewed in the environment, a cause was presupposed and then a supporting effect was looked for. If the supporting effect existed, then there was a supposed link between the original effect and the presupposed cause. As an example,

Effect	A woman is washing baby bottles and filling them with formula
Cause	The woman knows that in the future she will need to feed a hungry baby
Effect	A baby is present and the baby is the daughter of the woman

The cause of "the woman knows that in the future she will need to feed a hungry baby" is not directly related to "A woman is washing baby bottles and filling them with formula." However, if "A baby is present and the baby is the daughter of the woman," then the cause is probably correct.

Additional Cause

Additional cause refers to the existence of an effect whose magnitude cannot be explained by the existence of one observable effect. A pilot flying an airplane may be having problems maintaining a certain air speed because if a strong head wind. However, the situation could be magnified by having a defective engine.

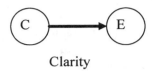

Clarity

Figure 4.31

Clarity

Clarity questions the cause-and-effect relationship. In other words, if clarity is lacking, the reader does not understand the relationship between the cause and the effect (Figure 4.31).

Note that the limitations of the policy analysis method is that it is only as effective as the knowledge of the person using it. It is understandable that a person experienced in running a production control department may have difficulty applying it to sales.

SUMMARY

Since most constraints to increasing the amount of throughput entering the company will be policy related, the adaptation of the TOC TP is a must. The TOC TP greatly enhances the ability of anyone to understand what must be done, how it can be accomplished and how to ensure that the solution does not generate more problems than it solves.

•   •   •

Nowhere is the need for policy analysis more evident than in making decisions.

•   •   •

STUDY QUESTIONS

1. What is the objective of the TOC thinking process?
2. Define the following terms and explain how they are used: undesirable effect (UDE), entity, intermediate objective, obstacle, injection, action.
3. What does the term "trimming negative branches" mean?
4. Define the term "categories of legitimate reservation."

5. List and define six categories of legitimate reservation. Show examples of each.
6. List the five tools used in the TOC thinking process and explain how each is used.
7. Present a specific conflict in your environment/life and use the TOC thinking tools to define and solve the problem.

Correcting the Decision Process

Possibly the greatest impact of any company's activities will come through being able to make consistently valid decisions. A sound decision-making process yields major and immediate benefits for the profitability of any company. *How* a decision should be made has radically, but necessarily, changed. It has become the subject of major policy analysis issues.

OBJECTIVES

- To begin to understand what the impact of current decisions are on the profitability of the company.
- To begin the process of understanding how decisions should be made.

• • •

The secret to making good decisions is in being able to predict the impact of the decision on Throughput, Inventory and Operating Expense. It is the environment and the limitations of the system which dictates the solution.

• • •

Not every decision a manufacturer will face can possibly be covered in one book. However, the following discussion provides insight into how decisions *should* be made.

QUALITY COST

To address Operating Expense, one can use the focusing mechanism of quality cost. Conventionally, it is necessary to know the magnitude of the impact of an occurrence on cost. The key indicators, therefore, are *cost* and the *magnitude* or *frequency* of occurrence. Cost can be related in a multitude of terms: internal failure costs such as scrap and rework, external failure costs such as warranty charges, appraisal costs such as inspection and test, and prevention costs such as process planning and control or education. There can be, quite literally, hundreds of these occurrences. To focus on only the most important issues, costs are arranged based on their impact. In a Pareto analysis, 80% of the costs are created by 20% of the cost drivers. The problem is that 20% can still represent a large number, and a correction in any one of these costs often results in only a very small improvement in profitability. In addition, the mechanisms for analyzing cost information can lead to corrections that have no impact on Operating Expense. Two excellent examples of problems encountered while attempting to use quality cost as a focusing mechanism are rework and scrap.

Rework

In Figure 5.1 there are two resources, A and B. Resource A feeds resource B. Each has a rework cost in the form of additional labor associated with a specific occurrence. For resource A the cost is $10 labor and for resource B it is $20 labor per occurrence. Resource A's frequency rate is 100 per week, and resource B's is 200. The total quality costs for rework are $1000 and $4000, respectively.

In this way, it is very easy to focus efforts. Traditional TQM approaches would require concentrating on the core cause for resource B's problem, which obviously has a greater impact because of the $3000 additional expense. However, in order to review this issue from a global perspective the concept of resource load must be added.

What is the financial impact if the "cost" is labor in a resource that has excess capacity to deal with the additional work? If resource B is loaded to 50% capacity and resource A is loaded to 100% capacity, which is more important to solve? What is the impact of solving a rework problem in resource A or B? These in turn

Cost		Freq		Total Cost
$10	x	100	=	$1,000
$20	x	200	=	$4,000

Figure 5.1 Computing quality cost.

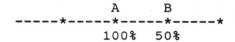

Figure 5.2 The impact of product flow.

raise some other interesting questions that must be answered in order to solve the problem.

- If the problem is solved at resource B, will the ability to ship additional product go up, thereby increasing Throughput?
- Will Operating Expense be reduced?
- Will Inventory go down?

Figure 5.2 helps one to understand the problem a little better. Resource A feeds resource B, so the actual output from resource B is regulated by the capability of A. Resource B has an additional 50% capacity that is going unutilized and can easily overcome any delays in shipping. So, by fixing resource B's rework problem, no additional products would have to be shipped. Throughput would remain level.

The next question is how much will operating expense be reduced? The immediate assumption is that it will be reduced by the amount of the rework cost, or $4000. However, from a global perspective, Operating Expense is caused by salaries, the rent for the facility and other things such as utilities and insurance. Will any of the cost drivers for Operating Expense go down? Since the worker is probably being paid for a 40-hour work week and not for work performed, this seems unlikely. Will the worker be laid of if the problem is fixed? This seems unlikely as well. From a global perspective, the $4000 savings for fixing resource B is a mirage.

The final question is how much of a decrease in inventory will there be? The question might also be asked whether the delay caused by resource B's rework problem has actually resulted in an increase in inventory in the first place. Since resource B exists after the constraint, the amount of inventory that would exist between the constraint and the end of the process is regulated by resource A. To result in an increase in inventory, a delay in shipment would have to occur. The additional 50% capacity at resource B should ensure that due dates are actually met.

If resource B were fixed, Throughput, Inventory and Operating Expense would still remain level. However, if any money is spent trying to solve this problem, Operating Expense would go up without a corresponding increase in Throughput, causing profitability to decline.

Now the same questions must be addressed to the global impact of fixing resource A. Since resource A is the governing factor for creating Throughput, and has a limited availability, any increase in A's availability will increase the amount of Throughput being generated. Unless resource A is working overtime, Operating Expense will not decline for the same reasons as discussed with resource B. Would inventory decline?

Whether or not inventory would increase or decrease depends on the rate of consumption as well as the rate of release of raw material into the system. Increasing capacity on the constraining resource would increase consumption. If the rate of release of material is not increased to keep pace, inventory will decline and Throughput will be threatened.

In short, fixing resource A would result in an increase in profit, while fixing resource B would result in no improvement or a possible decline in the bottom line—and yet traditional quality cost approaches would focus on resource B.

Scrap

Figure 5.3 represents two parts, each with a certain scrap ratio. Part A consists of $50 material $50 labor and $100 overhead. Part B consists of $25 material, $25 labor and $75 overhead. Material is the cost of the raw material needed to make the part. Labor is the expense of the direct labor expended by factory workers. Overhead represents costs that are not directly related to the labor being expended to produce the part, such as engineering expenses or rent for the facility. In traditional accounting, these are the basis for creating the standard cost. The profit margin is determined by subtracting the standard cost from the sales price.

Part A is an expensive part which sells for more money and has a greater profit margin. Part B is a less expensive part. The scrap rates are 20% for part A and 10% for part B. Each part is scrapped at the end of the operation.

Under the quality cost or standard cost system, whenever part A is scrapped, the loss is $200 standard cost. Whenever part B is scrapped, the loss is $125 standard cost. Obviously, it would be prudent to fix the scrap on part A. It loses more standard cost, has a higher scrap ratio, a higher sales price and a higher profit

Part	Mat.	Labor	Over Head	Standard Cost	Sales Price	Prof. Margin	Percent Scrap
A	50	50	100	200	300	100	20%
B	25	25	75	125	200	75	10%

Figure 5.3 Computing the impact of scrap.

margin. However, as in the previous example, there are other things to consider. What is the financial impact if part A is being made using resources that have excess capacity? What is the financial impact if part B is being made using resources that are scheduled to 100% capacity. In part A each time a part is scrapped the loss is the cost of raw material alone. There is excess capacity to replace all but the raw material. In part B each time a part is lost, the entire value of the sales order is lost. It is being made using resources that cannot donate any additional time to make up the difference.

Sometimes the alternatives are not quite that easy. What would be the impact if the volume of sales for part A were 100 pieces and for part B 40 pieces per month?

```
                                   Scrap   total
            QTY    Loss             rate    loss
     A     100 x  $50  =  $5,000 x  .20  = $1,000
     B      40 x $200  =  $8,000 x  .10  = $   800
```

In part A the loss is $1000 per month; in part B the loss is $800 per month. But with part B, because it is being made using resources that are scheduled to 100% capacity and cannot make up any loss, the loss also includes future potential for sales by not delivering on time. This may prove to be the biggest problem.

One primary tenet in TQM is that those employees who create an improvement should not suffer layoffs. Regardless of how streamlined the operation becomes, if sales do not go up and Operating Expense is not lowered, how can it be said that an improvement has occurred? If increasing the quality of the products and services does not result in an increase in sales and if lowering "quality cost" does not result in a decrease in Operating Expense, then there exists a major problem that cannot be ignored. In traditional TQM the focusing mechanisms must be modified.

COST ACCOUNTING

Cost accounting was developed around the turn of the century to "judge the impact of a local action on the bottom line" (Goldratt, 1990). The objective was to assign a value to an occurrence and to judge its impact on the bottom line in the form of cost. Over time it has proved to be inadequate because circumstances of its development have changed considerably. Labor is no longer totally variable, and overhead is now a considerably larger portion of cost. The advent of activity-based cost accounting (ABC) supports this observation. However, the same basic assumption that the secret to making good decisions is based on cost drivers has come into question. Even the creators of ABC are now having second thoughts. Dr. Robert Kaplan of Harvard University, one of the chief proponents of ABC, has said that ABC cannot be used to make decisions.

ACTIVITY-BASED COST ACCOUNTING (ABC)

The objectives of ABC are to determine (1) what the real cost of goods sold is so that a healthy product mix can be established for the sales force and (2) which products are absorbing the most overhead activity so that non–value added activity reduction programs can be properly focused. ABC recognizes that many activities are not related to volume, as in traditional cost accounting, and assumes that activities consume resources and products consume activities. However, when examined from a global perspective, ABC has some major flaws in trying to accomplish its objectives.

Applying ABC

Figure 5.4 represents the traditional cost accounting model with some additional features. Two products are shown with specific standard cost information along with the sales price and profit margins for each. Also included are the quantity sold and Throughput being generated by each product. (Remember that Throughput is defined as the sales price minus raw material. In this case, it has been multiplied by the quantity sold to determine the total Throughput generated by each part.) Operating Expense is later subtracted to determine the net profit. The quantity sold for each product is the ceiling for the market currently being served.

In this model, products A and B have the same standard cost but differing sales prices. Since product B sells for more, it is identified as the more profitable product while A is identified as the less profitable. In ABC, an attempt is made to re-allocate overhead, activities so that "real costs" can be determined. If product B

	Mat.	Labor	Over Head	Standard Cost	Sales Price	Profit Margin	Quantity Sold	Thruput Gen.
A	80	60	120	260	300	40	100	22,000
B	80	60	120	260	310	50	200	46,000
								68,000
						Oper. Exp.		−60,000
						Net Profit		8,000

Figure 5.4 Applying ABC.

were a new product and therefore absorbed more overhead the cost allocation model might be changed to look like Figure 5.5. Part B, which requires the activities of 10 engineers, now absorbs 75% of the overhead allocation for both parts.

The new overhead allocation model shows that product A is now the more profitable and product B is now the less profitable.

Addressing The Impact

In Figure 5.5, notice that while profit margins have changed, Operating Expense as well as the amount of Throughput being generated have not changed as a result of the re-allocation or manipulation of data. This would require that action be taken based on the information presented. So, from a global perspective, Net Profit remains the same. The two important questions are what kind of action will this information create, and what will be the impact on the profitability of the company? There are two immediate possibilities:

- The elimination of product B
- The reduction or elimination of non–value added activities associated with B

In addressing the first issue, what is the impact on the profitability of the company if product B is eliminated because it seems unprofitable? If product B is eliminated, so will the $46,000 worth of Throughput generated, leaving product A to absorb all Operating Expense. Unless Operating Expenses is reduced the resulting net profit will be a minus (loss of) $38,000 ($22,000 − $60,000).

In any attempt to reduce Operating Expense, it is highly unlikely that the total overhead associated with product B will be eliminated along with the product. If 10 engineers spent 75% of their time working on product B, 7.5 engineers could be eliminated. But the remaining 3 (rounded up from 2.5) engineers would probably need to stay to support product A. Along with the expenses of the

	Mat.	Lab.	Ovr. Hd.	Std Cost	Sales Price	Prof. Mar.	Quan. Sold	Thruput Gen.
A	80	60	60	200	300	100	100	22,000
B	80	60	180	320	310	−10	200	46,000
								68,000
								−60,000
								8,000

Figure 5.5 Adjusting for overhead application.

engineers, the expenses originating from the rent of the manufacturing facility and the managers and supervisors who work there will probably not be eliminated either. To maintain the same net profit at $8000, operating expenses must drop from $60,000 to $14,000. Few companies can manage a 76% drop in Operating Expense!

Additionally, if the limitation to increasing the cash being generated through sales is that there is no more market available, and if 70% of the engineers were eliminated, who would design the new products required to be able to increase cash coming into the company?

In addressing the second issue of focusing on eliminating the non–value added activities associated with product B, can an effective program be created that will have a direct impact on the reduction of Operating Expense or the increase of Throughput? Not likely. There are four issues to consider:

- The elimination of non–value added activities may have no direct link to the reduction of Operating Expense.
- At some point the elimination of non–value added activities may threaten those activities that support the generation of Throughput.
- Time spent concentrating on non–value added programs will distract from initiatives that may have a greater impact on the profitability of the corporation
- Reducing non-value overhead activities associated with product B will not result in the Exploitation or Elevation of the constraint (the market, in this case), and therefore Throughput will not increase.

Addressing the Seven Wastes

One method for concentrating on the elimination of non–value added activities is eliminating the "seven wastes":

- *Excess wait time* — time the operator spends waiting for a machine to process while he or she is watching. The thought is that the operator could be doing something else while waiting.
- *Excess transportation* — the unnecessary distance a part must travel for any purpose.
- *Excess processing* — any additional effort used to produce a product.
- *Excess inventory* — inventory not required by downstream operations
- *Excess motion* — motion that does not add value to the product.
- *Defects* — any defective part produced by any operation anywhere in the plant
- *Overproduction* — creating products that are not needed by downstream operations .

Removing the seven wastes is designed to improve the efficiency of activities associated with specific products. The thought is that if too much overhead activity is being spent on specific products, then the increased efficiency of those activities will reduce the amount of time spent and therefore will reduce cost. However, a reduction in the seven wastes, unless it is directly linked to the improvement of the constraint(s), may result only in increasing excess capacity available for any specific activity, and with respect to overhead-related labor, unless a reduction of the total work force is obtained, total Operating Expense will not go down. Even if the people are moved to another operation, they are still on the company payroll and represent Operating Expense. If Throughput does not go up or Operating Expense does not go down, productivity will not improve.

For a non–value added reduction program that focuses on the "seven wastes" to impact Operating Expense, employees would probably need to be eliminated. A process of ongoing improvement focusing on the reduction of non–value added activities, which would effectively impact profitability, may require the constant elimination of employees. What is the strategic impact of this situation? Excess capacity is a strategic weapon to be used to develop new market segments and helping the company grow. If eliminated, it can jeopardize overall corporate growth (see Chapter 16, "Facing the Strategic Issues.")

The Impact of Internal Capacity on ABC

Determining the most profitable product mix requires additional information not considered by ABC. For the products described, no mention has been made of any internal capacity limitations that might exist, and yet this is what controls the creation of Throughput coming into the company.

If the amount of capacity is limited for products A and B because of unavailability of a specific resource, a decision must be made to determine the most profitable product mix (Figure 5.6).

Resource 124, because of a lack of available capacity, places a limitation on the total number of A and B that can be produced. Using ABC, in the example presented, the decision would be to maximize the selling of A over B. If so, the production of 100 A would use 3000 minutes of the available time on resource 124. The remaining 2400 minutes would be left to produce 160 units of B. Figure 5.7 shows the resulting net profit.

While A seems to be the more profitable product under ABC, after reviewing the route file for both products, it is discovered that more of part B can be created in less time, resulting in more Throughput being generated. If the decision was based on maximizing profitability, given current limitations, part B would be the more profitable product. Efforts should be made to supply the available market with all 200 units of B; whatever time is remaining should then be devoted to making A.

```
Routing                          Routing
Part A                           Part B

Res  Op.  Time                   Res  Op.  Time
123  10   10                     123  10   10
124  20   30                     124  20   15
125  30   20                     125  30   20
```

```
      Time       Demand
Res   Avail.    A      B     Total   Delta
123   7200     1000   2000   3000   +4200
124   5400     3000   3000   6000   -  600
125   7200     2000   4000   6000   +1200
```

Figure 5.6   Internal capacity limitations.

	Mat.	Lab.	Ovr. Hd.	Std Cost	Sales Price	Prof. Mar.	Quan. Sold	Thruput Gen.
A	80	60	60	200	300	100	100	22,000
B	80	60	180	320	310	-10	160	36,800
								58,800
								-60,000
								- 1,200

Figure 5.7   The impact of internal capacity on ABC.

	Mat.	Lab.	Ovr. Hd.	Std Cost	Sales Price	Prof. Mar.	Quan. Sold	Thruput Gen.
A	80	60	60	200	300	100	80	17,600
B	80	60	180	320	310	-10	200	46,000
								63,600
								-60,000
								3,600

The formula for determining the most profitable product mix is (Cash generated)/(Unit of the limiting resource).

```
                           A                  B
    Sales Price           300                310
    Raw Material           80                 80

    Cash Generated        220  = $7.3        230  = $15.3
    Constraint Time        30                 15
```

• • •

In the final analysis, from a global perspective, physical limitations overshadow any decision process. Neither the profitability of the company nor the understanding of where to focus improvements can be predicted by either traditional or activity-based accounting.

• • •

PRODUCT MIX

Exploiting the constraint also includes deciding what product mix is the most profitable. Whenever a limitation exists restricting the amount of product that can be produced, a decision must be made to chose one product over another so that profits can be maximized.

Figures 5.8–5.11 represent a decision process involving the profitability of a company. In Figure 5.8, products AFG and DEF have a market potential of 70 and 30 pieces, respectively.

AFG is sold for $120 and DEF is sold for $140. There are four resources, A, B, C, and D. Each resource has available 2000 minutes during the week. The Operating Expense is $7000. Bills of material and routings have been combined, forming a network of processes beginning with gating operations where raw material enters, and proceeding upward until the finished products are complete (at the top). Raw material cost is presented in circles. Operations are performed in the amount of time and at the resources indicated in each box. To make part AFG requires that $35 raw material be used in the first operation (at the bottom) at resources D and C. Resource D takes 20 minutes to perform the first operation on the left leg of the net. This operation feeds the next operation to be completed at resource B. The final assembly for AFG takes place at resource A. The middle leg feeds both products. The right leg feeds product DEF. The initial constraint is found to be at resource C (Figure 5.9).

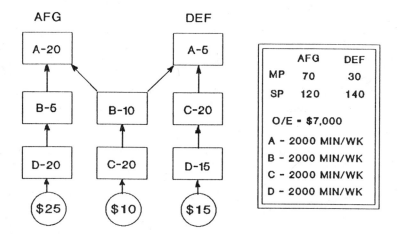

Figure 5.8 The decision process.

RESOURCES

PROD.	A	B	C	D
AFG	1400	1050	1400	1400
DEF	150	300	1200	450
TOTAL	1550	1350	2600	1850

Figure 5.9 Identifying the constraint.

The labor to produce 70 AFGs and 30 DEFs is 2600 minutes—600 more than the available 2000 minutes per resource.

The Traditional Approach

Because there is a limitation in the amount of time that can be used from resource C, in order to maximize profitability, a decision must be made concerning those products that will bring the most amount of profit for the given limitation. Traditionally, the decision process is based on the concept of profit margin resulting from the formula sales price minus the cost of goods sold. Product DEF is considered the most profitable. It has a sales price of $140 and a raw material cost of $25, while the required labor to produce the part is lower than that required for product AFG (Figure 5.10).

	AFG	DEF
PRICE	$120	$140
MATERIAL	$35	$25
PROFIT	$ 85	$115
LABOR	75MIN	70MIN

```
Throughput = 30 DEF(140 - 25) =   3,450
Throughput = 40 AFG(120 - 35) =   3,400
                                  6,850
                      O/E =      (7,000)
             NET PROFIT = $  -   150
```

Figure 5.10 The traditional approach.

Conventional wisdom would place priority on selling all 30 DEFs, creating $3450 ($140 sales price minus $25 raw material, multiplied by a quantity of 30) in Throughput and using 1200 minutes of resource C's time. Any remaining constraint availability (2000 − 1200 = 800 minutes) would then be used to make AFGs. Each AFG part takes 20 minutes to produce on resource C resulting in a total of 40 AFGs being made and generating $3400 [(120 − 35) × 40] for a total Throughput generation of $6850. After subtracting $7000 in operating expense, the resulting profit is −$150.

Understanding the Impact of Internal Limitations

The question that should always be asked is not what is the most profitable product, but what is the maximum amount of money that can be made given the limitations of the system? This is determined by the amount of Throughput generated per unit of the constraint. To determine the most profitable product mix, considering the limitations provided, the amount of Throughput being generated (sales price minus raw material) is divided by the amount of time used by the limiting resource to create it. In this case, the Throughput per unit of the constraint for AFG is ($120 − $35) divided by 20 minutes of constraint time, or $4.25 per minute. The Throughput per unit of the constraint for DEF is ($140 − $25) divided by 40 minutes of constraint time, or $2.88 per minute.

Throughput Generated	85	= $4.25
Constraint Time Used	20	

Throughput Generated	115	= $2.88
Constraint Time Used	40	

Throughput = 70 AFG(120 - 35) = $5,950
Throughput = 15 DEF(140 - 25) = 1,725
                                 ─────────
                                   7,675
                      O/E =   (7,000)
         NET PROFIT = $ 675

Figure 5.11 Constraint-based approach.

Using the same information but changing the decision process has a much different result. The amount of labor required from the constraint is much less for each unit of the AFG product. The revenue generated per unit of the constraint is higher (Figure 5.11).

When comparing the Throughput per unit of the constraint, it becomes obvious that the most profitable decision would be to maximize the creation of product AFG over DEF—just the opposite of the cost solution. However, the resulting net profit is also greater using this solution.

The choice of approach has a major impact not only on which products must be pushed onto the market but also on how all the other functions within the organization are to perform, including purchasing, production, engineering, materials and quality. If the constraint were to be elevated by adding another 2000 minutes to resource C, every function within the organization would need to reevaluate its priorities. If resource D becomes the constraint, AFG is no longer the preferred product to sell. The Throughput generated per unit of the constraint (resource D) now favors product DEF. All work schedules become subject to change, and any engineering, quality or production emphasis placed on maximizing Throughput is shifted to resource B.

MAKE/BUY

Any time markets decline and profitability begins to decrease companies invariably begin looking for ways to reduce cost. One cost reduction strategy involves the make/buy decision. The thought is that if a vendor can make a part

		Ovr	Std	Vendor
Mat	Lab	hd	Cost	Price
80	60	120	260	100

Figure 5.12 The make/buy decision.

cheaper and reduce the overall cost of products, the benefit will be carried to the bottom line. The traditional approach to answering this question is the cost method. The standard cost of the item in question is compared to the price to be charged by the vendor. Figure 5.12 represents a cost matrix showing material, labor and overhead for a given part. The standard cost is shown to be $260. The vendor's price for the part is $100. The obvious choice is to buy the part from the vendor.

However, as the secret to making valid decisions lies in understanding the impact on Throughput Inventory and Operating Expense, this decision will require further analysis. There are two basic considerations.

- If a part is currently being made outside the plant by a vendor, what will be the impact if it is now brought inside the facility?
- If a part is currently being made inside the plant, what will be the impact if it is now made at an outside vendor?

Like most cost-based decisions, the traditional process uses only cost as the primary factor and does not take into consideration internal resource capability and its impact. If a part is being made internally, what will be the impact on the ability to protect and create Throughput. There are three conditions that need expanding. They involve use of:

- Excess capacity
- Protective capacity of non-constraint resources
- Constraint time

Using Excess Capacity

Under those conditions where the part is to be built using resources that have excess capacity, the actual cost is only the cost of raw material. If the part is made internally, no extra personnel would need to be hired, so Operating Expense would not go up. The actual cost of the part being made internally (Figure 5.12) would be $80, compared with the vendors $100.

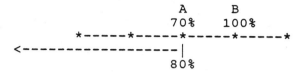

Figure 5.13   The impact of capacity on make/buy.

Using Protective Capacity

If the part were to be made using resources that are marginal in their ability to protect the constraint, adding additional labor demand would reduce the amount of protective capacity available.

Figure 5.13 shows the impact of using protective capacity to produce the part in question. Resource A is a non-constraint, which is scheduled at 70% capacity and is feeding the constraint, resource B. Material which was previously obtained from the vendor is now being brought internally and will absorb time from resource A. Resource A's load is now increased from 70% to 80%. This increase in load reduces the probability that A will be able to deliver to B on time.   To ensure that B is fully loaded, there are two choices to consider. Either material will need to be released earlier, resulting in an increase in Inventory and the Operating Expense associated with it, or additional Operating Expense will need to be used in the form of overtime. How much additional Inventory or Operating Expense there will be can be determined only after an analysis of what new limitations would be created.   This analysis may point to the creation of a secondary constraint.

If a part is being produced internally at resources that are marginally able to protect the constraint, to buy it instead would produce an increase in the amount of protective capacity available, resulting in Inventory or Operating Expense being reduced. It would also mean an increase in Operating Expense to cover the cost of the product arriving from the vendor.

The increase in Operating Expense would be by the amount of the order from the vendor. The reduction in cost could be determined only after physical analysis to determine the impact of the decreased load.

Using Constraint Time

Determining whether a part should or should not absorb constraint time can have several implications.

- Adding more constraint time for a particular product will reduce the amount of Throughput generated per unit of the constraint for that product
- Adding more constraint time for a particular part will impact the amount of time available to produce other products
- Reducing the amount of constraint time being absorbed by buying parts will increase the amount of Throughput generated but may result in an increase in Operating Expense
- Increasing the amount of constraint time by making a part will result in a decrease in Operating Expense originating from the vendor but may result in an increase in Operating Expense at the constraint

Product	A	B	C
Sales Price	$100	$150	$200
Raw Material	50	75	100
Throughput Created	$50	$75	$100
Constraint Time-Minutes	15	30	45
Throughput Per Min	3.33	2.50	2.23
Additional Time From New Part	15		
New Throughput Per Min	1.66	2.50	2.23

Figure 5.14 Adjusting the make/buy decision.

Adding demand requirements at constraint resources by making a part internally and increasing the amount of time a particular product absorbs may result in a change in product mix. When comparing Throughput per unit of the constraint to other products, a product that has had a relatively high ratio may now appear at the other end of the spectrum creating a low Throughput rate and resulting in a change of sales incentive (Figure 5.14).

Products A, B and C create $50, $75 and $100, respectively, in Throughput per part and take 15, 30 and 45 minutes to do so, resulting in Throughput per minute of $3.33 for product A, $2.50 for product B and $2.23 for product C. If an additional 15 minutes is added to product A's time at the constraint, changing its Throughput ratio to $1.66 per minute, product A goes from being the most profitable product to the least. A new product mix and profitability strategy may need to be created. The impact on other products being produced at the constraint cannot be totally known unless a schedule is built that can model the event.

Estimating capacity for the constraint at 3600 minutes available time, marketing constraints for products A, B and C at 50 pieces each, and using the original labor requirements prior to bringing the previously purchased part inside the plant, the following Throughput projection is made (Figure 5.15).

The total Throughput that could be made at this facility is $9250. After bringing the part in-house from the vendor the projection is modified (Figure 5.16).

The impact has been the elimination of product A from the product mix in order to maximize profitability and a drop in Throughput of $1100. To increase the amount of time available and thereby increase Throughput, overtime might be required, also increasing Operating Expense.

To understand the impact of off-loading product from the constraint to the outside vendor would require a reversal of strategy. Sending the 15 minutes created by product A to the vendor would have increased Throughput from $8350

Product	Throughput Per Part	Quantity Product Made	Constraint Time Used	Total Throughput
A	$50	50	750 Min.	$2,500
B	$75	50	1500 Min.	$3,750
C	$100	30	1350 Min.	$3,000
			3600 Min.	$9,250

Figure 5.15 Projecting Throughput.

Product	Throughput Per Part	Quantity Product Made	Constraint Time Used	Total Throughput
A	$50	0	0 Min.	0
B	$75	50	1500 Min.	$3,750
C	$100	46	2100 Min.	$4,600
			3600 Min.	$8,350

Figure 5.16 Adjusting for the new product load.

to $9250 but would have resulted in having to pay for the part in Operating Expense.

Additional Considerations

One consideration not made thus far involves product control. A "sensitive" part—one that will impact production to a great extent if it is poorly made or if its availability is unpredictable—may best be made in-house, where its production can be better controlled, rather than depending on a vendor to supply it. Thus it is important to study the process of out-sourcing products from all angles, not just in terms of cost.

PRODUCT PRICING

Product pricing is a key issue in the strategic positioning of any company. While price is usually set by the market, the acceptance of any order must include the seller's approval. Price acceptance is usually a function of the predicted profit margin obtained from the algorithm Sales price minus Cost of goods sold. Figure 5.17 represents three possible market segments. The first rectangle is the current market being supported by the XYZ company, in which product AFG is being sold for $300 and has a standard cost of $260.

Management has been informed by the sales group that two additional markets are available if prices can be dropped. In the first market, prices must be dropped to $200, and in the second $100. Traditional cost accounting would place the losses for the $200-segment at $60 per part sold, and the $100-segment at $160. Most companies would understand immediately that a certain amount of overhead can be spread over the new order and so the orders for the $200 market would be grudgingly accepted. Since the price in the $100 market doesn't even

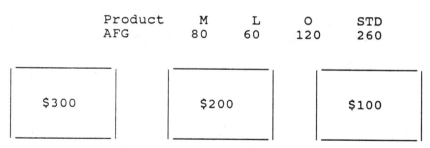

Figure 5.17 Product pricing in segmented markets.

cover raw material and labor costs, it would be rejected. However, as with the make/buy decision, the impact based on a cost matrix is totally unknown. If the parts for the new market segments were to be made at non-constraint resources, there would be no additional labor required to handle the new orders. Labor is a non-variable expense in that unless overtime were expended or personnel hired, payroll would remain at 40 hrs per employee per week. Figured this way, the profit for the $200 market would be $120 per part, and for the $100 market, $20 per part.

Price Acceptance

In segmented markets, for a company using non-constraint resources, the price should be the highest price that can be obtained above the price for raw material. As in the make/buy decision, any new orders may threaten protective capacity. If a part is made using resources for which additional protection would be required, inventories would increase.

If the part were to be made at the constraint, the amount of Throughput per unit of the constraint would need to be compared to what is currently being produced and, if found to be less, the order should not be taken unless other circumstances dictate. If the $300- and $200-segments were creating an internal constraint, it is obvious that taking an order from the $100-segment would mean a loss of an order from either of the other two segments. To properly segment the market:

- The sale of a product in one market segment should not negatively impact the sale of a product in another market segment
- Each market segment must use the same resources
- The segments should be flexible so demand in one market is down, the company will still have adequate business from the other segments

In order to accept an order, the sales representative must know:

- How much each unit of the constraint is being sold for now
- How much of the constraint will be absorbed by the product in question
- What the customer is willing to pay
- Whether the order will impact non-constraints negatively

LOT SIZING

The determination of proper lot sizes has a major impact on the overall productivity of the company. If lot sizes are too large, inventories and lead times begin to grow and quality decreases owing to a lack of visibility. Products begin competing for sufficient production time at resources with limited capacity, causing Throughput to be blocked. If lot sizes are too small, limited resource capacity is spent supporting setup operations and not production.

The Traditional Approach

The traditional approach was to use the economic order quantity (EOQ), which balanced setup cost with Inventory carrying cost. A larger batch resulted in fewer setups and therefore lower setup cost but higher carrying cost because of an increase in Inventory. A smaller batch resulted in more setups and therefore a higher setup cost but lower Inventory carrying cost.

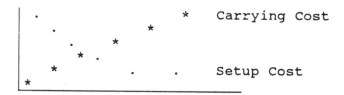

This approach looks great on paper but immediately falls apart in reality. First, it assumes that the setup cost for all resources is the same. However, a resource that has excess capacity will incur no additional expense unless additional setups cause protective capacity to be used. Increased setups at those resources with limited capacity will threaten the creation of Throughput. Increased setups at the primary constraint will cause Throughput to decline immediately. Second, it assumes that the primary costs associated with increased lot sizes are an increase in overhead costs such as interest rates, materials handling, storage and insurance, whereas the largest problem is the impact on restricting the creation of Throughput. Third, it assumes that the batch size will always remain the same throughout the production process. However, the size of the process and transfer batches may be very different. A group technology cell may process in batches of 100 but it will be transferred between machines in a quantity of 1.

In JIT, the optimal lot size has been described as the quantity of 1. The reasons given most often for this are:

- Increase visibility
- Better synchronize production with market demands
- Increase quality
- Increase flexibility
- Decrease lead time

While these are worthwhile attributes for any manufacturing facility, unless a global perspective is gained through a review of the impact of any lot sizing policy based on Throughput, Inventory and Operating Expense, there will be a misinterpretation of the requirements. It must be understood that because the manufacturing environment is dynamic, lot sizes are dynamic as well. In other words there is no optimal lot size for all situations; there is only the impact certain lot sizes will have, given the current situation with respect to available capacity. So determining the correct lot size means determining what the relationships between the resources at which the lot is to be processed and the time at which it is to be processed. This issue can be easily addressed when reviewing the impact of setup savings on resources.

The Impact of Setup Savings

In setup savings, the objective is to combine two orders so that one setup can be split between them, thereby maximizing the amount of production time available.

In Figure 5.18, to maximize the availability of production time at a constraint resource, product D has been pushed up in the schedule so that products A and D can be processed together. As a result, products B and C were pushed out and the total production time increased by the amount of setup saved.

Maximizing production time on a non-constraint operation is usually a waste of time. The only impact may be to increase excess capacity, as seen in Chapter 3, on setup reduction. But, at a constraint resource, setup savings, like setup reduction, has a positive impact on the amount of Throughput created. This is easy to visualize. However, maximizing production at the constraining resource

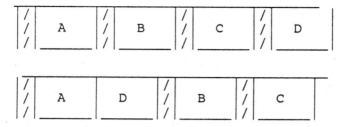

Figure 5.18 Setup savings.

may have a negative effect on those resources that are loaded to near-constraint levels.

In Figure 5.19 resource A is a non-constraint resource, but it is loaded to 80% capacity and feeds resource B, which is loaded to 100% and is the primary constraint. Resource B's production time and setup time are 80% and 20%, respectively. Setup savings is performed, increasing the amount of production performed at resource B.

If setup savings is performed and production is maximized at resource B, unless the amount of setup savings has the same productive impact on A, the load at resource A will increase. The laws governing probability and statistical fluctuation will begin to affect A's ability to deliver to B, threatening Throughput. To offset this, material is released earlier, causing Inventory to increase. There is a limit to the impact earlier release will have. Parts cannot be released earlier than time zero. So, to offset for an inability to protect the constraint, overtime will be spent, driving Operating Expense upward.

These same problems will occur for resources that have operations that follow the constraint. Figure 5.20 illustrates that when the productive capacity of the

```
                    A        B                       Prod  S/U
                   80%      100%      Before          80    20
       *-----*-----*-----*-----*
       <------------- 90%                After          90    10
```

Figure 5.19 The impact of setup savings on near-constraint resources prior to the constraint.

```
        A       B                      Prod S/U
       100%     80%        Before       80   20
*------*------*------*------*
                90% -----> After        90   10
```

Figure 5.20   The impact of setup savings on near-constraint resources after the constraint.

constraint is increased, protective capacity at the non-constraints that follow the constraint is decreased. If the impact is severe enough, sales order delivery is threatened.

Notice that in Figure 5.20 orders B and C have been pushed out. This will result in lowering the amount of protection available for ensuring that the sales order is shipped on time. Unless the amount of protective capacity is known at the resources between the constraint and shipping, the impact on Throughput, Inventory and Operating Expense cannot be known.

The Small-Lot Strategy

The objective of the small-lot-size strategy is to increase visibility by allowing each part to be observed shortly after production and to decrease lead times. Although reducing the lot size creates greater visibility, more time is used in setting up an operation. If the resource at which the small lot is used is a constraint, then Throughput will go down, owing to a decrease in production time available. Unless the impact of setup can be eliminated through a setup reduction program, the loss can be substantial. However, setups produce side effects, creating disruptions in the flow of products and increasing the opportunity for error.

Conclusion

The large batch maximizes the utilization of a given machine but extends lead times and increases inventories. The small batch increases visibility and decreases lead times but lowers the amount production at constraining resources by increasing the amount of setups required. The ideal situation would support the desired characteristics of both large and small batches but would be subject to modification based on the dynamics of the current environment. Lead time and visibility are greatly enhanced by using a transfer batch of 1.   Whereas productivity at constraint and near-constraint resources is enhanced by using large production batches.

The size of the production batch can be determined only after a load is placed at all resources for the given schedule and its impact on those resources that have a marginal capability to produce is determined. As a rule of thumb, the best process batch size to start with is the actual customer demand. Adjustments can be made upward by combining customer requirements, if necessary, to maximize Throughput. Transfer batches should be made as small as possible to keep lead times low and visibility high. The end result will be to maximize Throughput while keeping Inventory and Operating Expenses low.

COST JUSTIFICATION

The traditional justification for an improvement is usually based on the amount of labor saved per unit of product being produced multiplied by the labor rate and the number of parts to be produced. However, when reviewed from a global perspective, this method begins to fall apart. In Figure 5.21, as part of a cycle time reduction program, an engineer offers an improvement to resource B on the left leg of the net by cutting the amount of required labor by 50% at a cost of $3000. The productivity model lends some insight into what is important.

$$\text{Productivity} = \frac{\underline{\text{Throughput}}}{\text{Operating Expense}}$$

Resource B, being a non-constraint resource, does not control the creation of Throughput. It already has excess capacity. When the $3000 is spent, no additional money will be coming into the company. Since reducing the amount of labor required will not erase the need for resource B, payroll will not be reduced. Operating Expense will go up by the $3000, while Throughput will remain level, forcing productivity to decline. If the $3000 is spent to reduce the amount of labor required at resource C, then the amount of product being produced will increase along with the Throughput being generated. Operating Expense will go up, but so will Throughput. The key issue now is whether the Throughput being generated will pay for the investment in less time than resorting to another option.

Cost justification does not always lead to the intended results. A more effective method would be to Throughput justify.

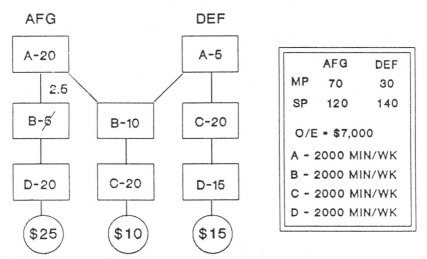

Figure 5.21 Cost justification—the decision process.

GLOBAL VERSUS LOCAL MEASUREMENTS

In each of the decision processes given, the attempt to arrive at a logical conclusion is made from a local perspective. However, when viewed from a global perspective (Throughput, Inventory and Operating Expense), the traditional decision processes often cease to make sense. Before any real long-term improvement can be attempted, these issues must be addressed. Many companies spend enormous resources trying to reduce cost of goods sold only to find that Throughput has not gone up and Inventory as well as Operating Expense have not gone down—and that profitability has not improved.

STUDY QUESTIONS

1. When was cost accounting developed, and why is it a poor method of decision support?
2. Explain the concept of quality cost, and why it does not work effectively to focus the improvement process.
3. Define activity-based cost and give examples of why it is not an effective decision model.
4. Under TOC, what measurement is used to support the concept of product mix, and how is it used?

5. Why do Inventory and Operating Expense exist?
6. Define the concepts of excess, protective and productive capacity, and explain the impact of each on the manufacturing company.
7. What is the benefit of market segmentation, and what are the rules for its use?
8. Explain the difference between a process batch and a transfer batch. What is the best way to determine the best size of each to use?
9. What is meant by the concept of Throughput justification?

Improving on Just-in-Time:
The DBR Process

This chapter introduces a superior method for factory scheduling/management that embraces more readily the way in that resources interface to produce a more effective improvement process—the drum-buffer-rope (DBR) process.

OBJECTIVES

- To understand how to create the drum for maximizing constraint utilization
- To understand what data is required and how it should be manipulated
- To understand how to create and implement a more effective im-provement process that has a direct link to the absolute measurements
- To understand how the factory schedule is an integral part of the five-step improvement process

THE TOYOTA KANBAN SYSTEM

The kanban system starts with the customer and extends through the production facility to the vendor. Its objective is to control the level of inventory, reduce lead times and synchronize the factory as well as vendors with the market. It has been called a "pull" system because demands are pulled from downstream operations based on customer requirements. Resources are synchronized through the use of permission slips called kanban or cards. In the two-card system depicted in Figure 6.1, there are two types of cards: the production and the move. Material cannot be produced or moved unless a card granting permission is available. Material that has been converted or is to be converted is stored in the outbound or inbound side in containers of small lots. Whenever the succeeding operation has a demand

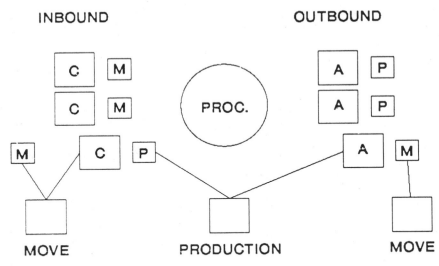

Figure 6.1 The two-card system.

placed on it requiring that parts be taken from the outbound side of the process, as the material is being carried away, the production card that has been attached is removed and used to secure parts from the inbound side so that replacement cards can be made. At the outbound side of the process, the A parts are moved using the M card and the P card is used to obtain the C parts for processing. Without a P card, the operator cannot begin production and will remain idle. Inventories are not allowed to build as in traditional manufacturing. Lot sizes remain low, which keeps lead times low as well.

In the one-card system, the inbound and outbound sides of resources are combined so that resources share inventory location. Since material does not require a separate movement system, the move cards are eliminated and only the production cards are used. In a cardless system, action is taken simply by the absence or presence of material as a signal; no cards are used.

For the repetitive manufacturer, Just-in-Time (JIT) has created a major opportunity for reducing lead times and increasing productivity. However, there are some major problems that must be overcome.

Lack of Flexibility

Just-in-time lacks the flexibility required by most manufacturers and therefore its applicability is limited to the repetitive environment. In the one- or two-card system, Inventory is stored in the kanban squares at the container and is an amount to be used as buffers between operations and to shorten the time and distance traveled to and from the stock room to find replacement parts. Since the predictability of configuration is much lower for the job shop environment than for the repetitive manufacturer, a much larger Inventory must be maintained between operations for replenishing a larger number of configurations, making this alternative too expensive. Some manufacturers have combined MRP with JIT to provide the flexibility needed (synchronous MRP), but the results have not met expectations. Inventory in the kanban square as well as kanban must be replaced on a regular basis with material for the new configurations and the old material stored, producing a logistical nightmare, especially for large facilities.

Vulnerability

Since each resource is dependent on the succeeding resources for production signals, whenever there is a disruption in the flow due to quality problems, a lack of materials or problems in setup, the entire line is shut down. Operations upstream of any disruption will cease to get production signals from down-stream operations, while operations that appear after any disruption will be starved for parts.

Creation of Additional Inventory

Kanban, while it has been touted as a pull system in synchronizing the shop floor, is actually a push system that creates more Inventory than is needed. It pushes Inventory into the kanban square whether or not it is actually required. Just because a part is removed from the outbound side does not mean it needs to be replaced.

Disruptive Improvement Process and Lack of Focus

one method of creating an improvement in JIT is to remove Inventory from kanban squares until a resource is unable to fill the succeeding resource's demand in a sufficient amount of time, causing disruptions in the production process. The disruptions are corrected by decreasing the amount of time required to replace the Inventory. This reactive method causes the output of the facility to decline immediately and may not result in an overall increase in Throughput after the disruption has been repaired, because the long-term output of the facility may be governed by a different resource in the system in which the disruption has been manifested as a problem.

Those improvements that are focused on the reduction of waste at and between all operations as measured by cycle time may have an impact on the reduction of Inventory but will not have an impact on the overall output of a manufacturing facility until those resources that have limited ability to produce are improved.

Finally, JIT represents a poor implementation of the five-step improvement process. If the objective is to maximize the output of a facility, the resource that has the greatest limitation, and not the market, must govern the way in that the factory is scheduled. The creation of the schedule at the constraint represents a portion of the exploitation phase and may include efforts to maximize the constraint's ability to produce Throughput, thereby invalidating the schedule established by the market. The trick is to accomplish the reschedule in such a way that sales orders are not late. Just-in-Time avoids this issue.

Long Implementation

The implementation of JIT is a long-term and complex endeavor. It requires a more sophisticated and better trained work force than is generally found. The small-lot-size strategy requires that vendor, setup reduction and quality programs be very successful to prevent drastic fluctuations in output. Typical implementations take several years a tremendous commitment in time, resources and money.

THE DRUM-BUFFER-ROPE METHODOLOGY

The DBR methodology is a technique for developing a smooth, obtainable schedule for the plant and for maximizing and managing the productivity of a manufacturing facility from a global, not a local, perspective. It differs from other manufacturing techniques in that it concentrates on determining the relationships among resources in resolving conflicts to create a smooth flow of product and is applicable to all types of processes whether they are repetitive, process or job shop. Drum-buffer-rope also provides an improved method of focusing protection so that the impact of disturbances on smooth production flow can be minimized.

•   •   •

The DBR process was designed as a means of implementing the five-step process of continuous profit improvement and therefore represents a tremendous leap forward in managing the shop floor from a profitability perspective. It is distinguished by representing how a factory should be scheduled based on the Theory of Constraints.

•   •   •

The Drum

The drum is the schedule for the system's constraint(s) and represents a portion of the exploitation phase of the five-step improvement process. It is used to maximize the available time of the constraint and to create the Master Pro-duction Schedule (MPS). Like the bass drum in a marching band, it is the drumbeat of the manufacturing facility. All other resources produce in synchronization to the constraint's schedule.

In order to schedule the constraint, an attempt is made to place the start and stop times for each order on a time line so that two conditions are met:

- Enough protection is available to ensure that each sales order due date is met.
- No conflicts exist between orders attempting to occupy the same space at the same time.

While the second condition must be met to create a valid schedule, the first condition is subject to the results of the second. If time is not available, sales orders will be pushed out and due dates not met.

Additional considerations arise when secondary constraints begin to appear. These are resources that have been scheduled to near-capacity levels, and because of this, will have trouble meeting the demands of the primary constraint schedule. After the primary constraint has been scheduled, resources that are loaded to near-constraint levels must be protected to ensure that the schedule for the primary constraint can be met. Secondary constraint schedules must be built so that whatever time is available at the secondary constraints can be maximized. However, the secondary constraint schedule must consider the schedule already established for the constraint. So, when building the secondary constraint, an additional consideration must be added: there must be no conflicts between the primary and secondary constraint schedules.

Since the primary schedule has been set, the secondary schedule must attempt to schedule around it. If unsuccessful, a reschedule of the primary constraint must take place.

The Buffer

The buffer is a time mechanism used to allow for those things that will go wrong, and it determines the lead time for products from the gating operations. The buffer is equal to the processing time plus the setup time plus an estimate of the aggregated amount of protective time required to ensure that the product will get to the buffer origin when needed. There are three areas (buffer origins) that need protection:

- Shipping to ensure that parts are delivered to the customer on time

- The constraint to ensure maximum utilization of resource time
- Those assembly operations in that one leg of the process is fed by a constraint and the other is fed by non-constraints so that parts that have been processed at the constraint will not wait in the assembly operation before parts from other, non-constrained resources arrive.

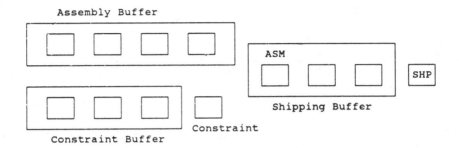

The Rope

The rope is the synchronization mechanism for the other resources and consists of the release schedule for all gating operations. Technically the rope is equal to the constraint schedule date minus the buffer time. The release of material determines the timing for parts being processed on the non-constraint resources.

APPLYING DRUM-BUFFER-ROPE

Establishing The Data Base

Figure 6.2 displays the sales order, product flow, inventory and resource data information used to compute the placement of orders. The sales order and re-source information is self-explanatory. However, the product flow, inventory and resource data information referred to as the Net will need some discussion. The top blocks in the Net represent the two products listed in the sales order file, AFG and XYZ. The larger, segmented blocks represent each part/ operation or station, the resource, and the amount of time required to process each part (similar to the product flow diagram discussed in Chapter 3). In the diagram, the first large block appearing at the top of the structure (in bold) is an assembly operation for part AFG, operation 10. It is shown as AFG/10 and takes ten minutes to process at resource A. Processing time appears just to the right of the resource information and above the part/operation. Part/operation ABC/10 appearing at the bottom of the left leg of product XYZ and (also in

Sales Order	Part	Quantity	Due Date
S/O123	AFG	20	154
S/O124	XYZ	10	154
S/O125	AFG	20	155
S/O126	AFG	20	155

The Net

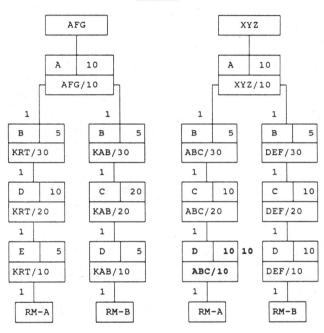

Resource Data

Resource	Capacity	Setup	Quantity
A	960	15	1
B	480	30	1
C	480	15	1
D	480	10	1
E	480	15	1

Buffer Data

Type	Length
Shipping	16 hours
Constraint	8 hours
Assembly	9 hours

Figure 6.2 The DBR data base.

bold) has 10 pieces of Inventory finished at that location and ready for processing at operation ABC/20. Inventory appears just outside and to the right of the block. The setup time is shown in the resource data appearing below the Net information.

Developing the Drum

Placing Orders on the Time Line

The first step in placing orders on a time line for creating the drum is to ensure that adequate protection exists between the sales order due date and the time an order is due to be completed at the constraint. To protect the shipping date and to determine the approximate time an order must cross the constraint, the shipping buffer is used. The date of the initial placement of an order will be equal to the sales order due date minus the amount of the buffer (aggregation of Murphy's Law).

Resource C has been tentatively identified as having the largest load, so developing the schedule will begin there.

- Each sales order is assumed due at the last hour of each 8 hour working day
- The shipping buffer is estimated at 16 hours, or 2 days
- To place each order on the time line for resource C, each sales order due date is used to begin the processing
- Current bill of material relationships and available Inventory above the constraint are established so that the load can be determined for each order crossing the constraint

The downward processing through the product structure begins with the farthest out sales order on the horizon: sales order S/O126, for product AFG, quantity of 20 and due on shop date 155. Since the schedule being created at this time is for resource C, only those parts/operations processed at resource C need to be considered.

Determining the Load

To determine the total demand of resource C's time created by S/O126, the total processing time for each part/operation is multiplied by the quantity required. Since there is no inventory available and since there is a one-to-one relationship between levels in the product structure, S/O126 will create demand for 20 pieces of KAB/20 at 20 minutes per piece, or 400 minutes. The 30-minute setup time requirement for the C resource is added, for a total of 430 minutes.

Determining the Completion Time

The initial placement of orders is determined by adding the buffer time to the sales order due date and time. The completion time for KAB/20 S/O126 is set at day 155, hour 8:00 (155/8:00) minus the 16-hour shipping buffer, or day 153, hour 8:00 (153/8:00).

Determining the Start Time

The start time for the order is set for 153/8:00 minus 430 minutes, or 153/0:50.

Completing the Initial Placement

The 430 minutes required for S/O125 part/operation KAB20 is also placed at 153/8:00. Each sales order is processed so that all parts/operations that cross the constraint are placed. The following figure shows the load placement for days 152 and 153 at resource C. Figure 6.3 is a graphical representation.

Sales Order	Ord. Due	Part/ Oper.	Proc. Qty.	Setup Time	Proc. Time	Start Time	End Time
S/O123	154	KAB/20	20	30	400	152/0:50	152/8:00
S/O124	154	ABC/20	10	30	100	152/5:50	152/8:00
S/O124	154	DEF/20	10	30	100	152/5:50	152/8:00
S/O125	155	KAB/20	20	30	400	153/0:50	153/8:00
S/O126	155	KAB/20	20	30	400	153/0:50	153/8:00

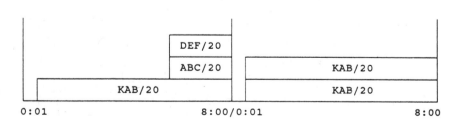

Figure 6.3 Placing orders on the time line.

Sales Order	Ord. Due	Part/ Oper.	Proc. Qty.	I N	Setup Time	Proc. Time	Start Time	End Time
S/O123	154	KAB/20	20		30	400	150/6:06	151/5:16
S/O124	154	ABC/20	10	10	30	100	151/5:17	151/7:27
S/O124	154	DEF/20	10		30	100	151/7:28	152/1:38
S/O125	155	KAB/20	20		30	400	152/1:39	153/0:49
S/O126	155	KAB/20	20		30	400	153/0:50	153/8:00

Figure 6.4 Leveling the load.

Leveling the Load

Obviously, the load must be leveled and sense that protection has been placed to the right. The only direction in which the load can be leveled is to the left. To create a level schedule, orders must be leveled beginning with the end of the horizon, and the start date and times must be placed end to end (Figure 6.4).

The first part/operation to be placed is KAB/20 for sales order S/O126. Its ending time is placed at day 155, hour 8:00 minus the 16-hour buffer, or 153/8:00. Its start time is 153/8:00 minus the 430-minute setup and processing time, or 153/0:50. Part/operation KAB/20 for S/O125 ends at 153/0:49 and begins at 152/1:39. This process continues until all orders have been placed on the time line for resource C.

Ensuring Constraint Utilization

Notice that part/operation KAB/20 for S/O123 has no Inventory available to begin processing. Since this situation will immediately invalidate the schedule, a decision must be made to process those parts/operations that have available Inventory. Since ABC/20 for S/O124 has Inventory, it will be processed first.

Sales Order	Ord. Due	Part/ Oper.	Proc. Qty.	I N	Setup Time	Proc. Time	Start Time	End Time
S/O124	154	ABC/20	10	10	30	100	150/6:06	151/0:16
S/O123	154	KAB/20	20		30	400	151/0:17	151/7:27
S/O124	154	DEF/20	10		30	100	151/7:28	152/1:38
S/O125	155	KAB/20	20		30	400	152/1:39	153/0:49
S/O126	155	KAB/20	20		30	400	153/0:50	153/8:00

The time from 150/6:06 until 151/0:16 should be spent processing Inventory that feeds the constraint based at the constraint's new schedule so that

Part/Operations

ABC20	KAB20	DEF20	KAB20	KAB20
150	151	152	152	153

0

Date/Time

Figure 6.5 Ensuring constraint utilization.

it does not run out of material. It should be noted that in DBR, material is allocated based on first come/first serve, while the schedule for the constraint is based on a finite setback methodology. Graphically, the new schedule looks like Figure 6.5.

Rescheduling for Time Zero

Notice that in the graphical schedule time, zero appears at the beginning of day 151. This means that the time available for day 150 does not actually exist; it occurred the day before, so a rescheduling must take place, pushing out the schedule of each order by the amount of time scheduled into day 150. Part/operation ABC/20's schedule has been placed 114 minutes into day 150. To reschedule, part/operation ABC/20 for S/O124 must be placed at time zero on day 151 and a schedule created, processing from the earliest time to the latest. Figure 6.6 shows the new schedule.

Notice that the original protection of 16 hours allotted for ensuring that orders arrive at shipping on time has been maintained for all but one order. Part/operation KAB/20 for S/O126 now has a scheduled completion date of 1 hours and 54 minutes past the 16-hour buffer. Since this is less than 50% of the total protection allowed, ordinarily, it should not be a problem. The schedule is considered to be on time.

If the scheduled completion date being over 50% of the buffer time were a problem, something would need to be done to gain more output from resource C

Sales Order	Ord. Due	Part/ Oper.	Proc. Qty.	I N	Setup Time	Proc. Time	Start Time	End Time
S/O124	154	ABC/20	10	10	30	100	151/0:00	151/2:10
S/O123	154	KAB/20	20		30	400	151/2:11	152/1:21
S/O124	154	DEF/20	10		30	100	152/1:22	152/3:32
S/O125	155	KAB/20	20		30	400	152/3:33	153/2:43
S/O126	155	KAB/20	20		30	400	153/2:44	154/1:54

Figure 6.6 Rescheduling for time zero.

such as setup savings, overtime or off-loading onto another resource.   If unsuccessful a rescheduling of the sales order may be required.

Developing the Rope

So far what has been accomplished is the schedule for the constraint. However, each operation must be synchronized to the demands of the constraint or the sales orders. The release schedule for raw materials into gating operations serves as the synchronizing mechanism.

In developing the rope, the release dates of raw materials into gating operations are also determined by the amount of protection necessary.

- For those items that go through the constraint, the drum schedule minus the constraint buffer equals the release date.
- For those items that do not go through the constraint, the sales order due date minus the shipping buffer is used.
- Whenever a part does not go through the constraint, but is used in assembly with parts that do, the assembly buffer designates the release date.

Figure 6.7 illustrates the release schedule and raw material demand to meet resource C's requirements from the above schedule.

Protecting the Constraint

In Figure 6.6, part/operation KAB/20 for S/O126 for a quantity of 20 is scheduled across the constraint at 153/2:44. The amount of protection required for the constraint is 8 hours, resulting in a raw material requirement for RM-B of 20 parts due at 152/2:44.

Raw Mat.	Quantity	Release Date
RM-B	20	151/0:00
RM-B	10	151/1:22
RM-B	20	151/3:33
RM-A	20	151/7:00
RM-B	20	152/2:44
RM-A	40	152/7:00

Figure 6.7 Developing the release schedule.

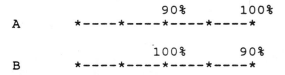

Figure 6.8   Adjusting for near-constraint resources.

Offsetting for Assembly Buffer Requirements

Demand for 20 units of RM-A is also created by S/O126 and is used in the non-constrained leg feeding the assembly operation AFG/10. Its timing is set for 155/8:00 minus 16 hours shipping buffer minus 9 hours assembly buffer, placing the release time at 152/7:00. Demand of RM-A for S/O125 is due also at 152/7:00 and has been combined with S/O126's RM-A demand. RM-B required on KAB/20 for S/O123 is due at 151/2:11 minus 8 hours, or 150/2:11. Since this is past time zero, the time is moved to 151/0:00.

Dynamic Buffering

Dynamic buffering is a method used to improve the buffering process so that overall buffer sizes can be shrunk, decreasing Inventory, and only for times when a non-constraint resource's protective capacity is threatened will an increase in buffer size and the resulting earlier release date of raw material be considered. Two conditions are required, as illustrated in Figure 6.8.

In example A, a resource that exists between the constraint and the release of material has been loaded to 90% as a result of the constraints schedule, limiting the amount of protective capacity necessary. The same condition exists in example B. Moreover, the resource with limited protective capacity exists between the constraint and the sales order. These are the conditions under which an increase in the amount of protection that is necessary will result in an earlier release of material or a rescheduling of the sales order so that additional capacity can be found. These conditions call for dynamic buffering.

To determine when dynamic buffering is appropriate, a comparison must be made between the amount of capacity available and the demand placed on each non-constraint resource being utilized by parts/operations that exist before and after the constraint schedule.

FOCUSING ON INVENTORY

Establishing the Basis

Throughput exists to create wealth. It is the mechanism by that money enters the company. Inventory and Operating Expense exists to cause the Throughput figure to be generated and also to protect it. Only once a rate of generating Throughput has been established for the weak links in the organization can the necessary level of Inventory and Operating Expense be determined.

Reducing Inventory requires that the blockers to reduction be identified. However, the objective is not just to reduce Inventory but also to protect the creation of Throughput. Throughput is created at the time of sale, so what must be protected is the delivery of the product to shipping. Those resources that have been loaded to the extent that they threaten the delivery of product to shipping must also be protected. Inventory is used to protect these deliveries. So the issue becomes not what is blocking the reduction of Inventory but what is blocking the reduction of *protective* Inventory. There are two issues:

- The amount of capacity at the non-constraints resources available to overcome those things that are going wrong
- The individual incidences that increase the amount of protection necessary

In Figure 6.9, resource B is loaded to 100%, and anything that goes wrong will result in an immediate loss of Throughput. Resource A is loaded to 80%, and if there is a problem, sometimes it will not affect Throughput and some-times it will by not delivering to B on time.

To solve resource A's problem additional capacity is necessary to protect Throughput. There are three ways of gaining additional capacity:

- To reduce the number of occurrences that use up the available protective capacity
- To obtain more of resource A from outside the company
- To release material earlier so that the demand is spread over more available capacity

Releasing materials earlier results in raising inventories. To reduce inven-

```
          A       B
         80%    100%
*-----*-----*-----*-----*
```

Figure 6.9 The impact of inventory.

tories, it is necessary to increase the amount of protective capacity available by eliminating occurrences that use protective capacity.

A prerequisite to reducing Inventory is an understanding of when something is supposed to arrive at those areas that are to be protected and to know whether or not it has arrived. Furthermore, knowing *why* something is late will help one begin to focus attention on those things that are blocking the on-time arrival. However, looking for something that is already late is not enough. What is really needed is an understanding of those actions that threaten to make something late before it happens. The secret to this is to look in front of the time something is supposed to arrive.

Buffer Management

Buffer management is a technique used to manage the amount of protection necessary and to focus improvements on those areas threatening the creation of Throughput and reducing the amount of protection required. An estimate of the amount of protection needed is used to determine the release date of material from the gating operation.

In Figure 6.10, a schedule for the constraint resource has been developed and a release date determined after estimating the amount of protection necessary at six hours. This time includes the time required to process the parts at each resource from the gating operation to the protected resource and the amount of time needed to guarantee that, most of the time, the parts will be delivered on time. In this case, a specific order was scheduled to begin production at the protected resource or "buffer origin" by 8:00. A 6-hour buffer required that parts be released at the gating operation at 2:00 in order to reach the protected resource by 8:00 so that a specific sales order could be filled on time.

The closer to the 8:00 deadline without the arrival of the material, the more threatened the buffer origin becomes. What is needed is a signal to designate when it would be time to find out what is wrong and to expedite so that Throughput is not jeopardized. In buffer management, the amount of time given to produce the product is divided into three equal time periods, or zones. Zone 1 is called the expedite zone. If the order does not arrive at the buffer origin before the beginning of zone 1, it is to be expedited. Zone 2 is the tracking zone. If the order has not arrived before the beginning of zone 2, it should be located to ensure that nothing is wrong.

```
Gating                Estimated Protection              Protected
Operation   2:00  |--------------------| 8:00  Resource
```

Figure 6.10 Estimating protection.

Z N	Sales Order	Ord. Due	Part/ Oper.	Proc. Qty.	INV	Setup Time	Proc. Time	Start Time	End Time
1	**8/O124**	**153**	**ABC/20**	**4**	**4**	**15**	**40**	**151/0:00**	**151/0:55**
1	**8/O124**	**153**	**DEF/20**	**4**	**4**	**15**	**40**	**151/0:56**	**151/1:51**
1	**8/O125**	**153**	**KAB/20**	**4**	**4**	**15**	**80**	**151/0:52**	**151/3:26**
1	S/O126	153	KAB/20	4	0	15	80	151/3:27	151/5:02
1	**8/O127**	**153**	**ABC/20**	**4**	**4**	**15**	**40**	**151/5:03**	**151/5:58**
2	**8/O127**	**153**	**DEF/20**	**4**	**4**	**15**	**40**	**151/5:59**	**151/6:54**
2	S/O128	154	KAB/20	5	1	15	100	151/6:55	152/0:50
2	**8/O129**	**154**	**ABC/20**	**5**	**5**	**15**	**50**	**152/0:51**	**152/1:56**
2	S/O129	154	DEF/20	5	0	15	50	152/1:57	152/3:02
3	S/O130	154	KAB/20	4	0	15	80	152/3:03	152/4:38
3	S/O131	154	ABC/20	5	0	15	50	152/4:39	152/5:44
3	S/O131	154	DEF/20	5	0	15	50	152/5:45	152/6:50
3	S/O132	155	KAB/20	4	0	15	80	152/6:51	153/0:26

Figure 6.11 The buffer report.

Buffer Reporting

Figure 6.11 shows a buffer report for a constraint where the buffer is equal to 16 hours. Each zone is 5 hours and 20 minutes. Zone 1 is from 151/0:00 to 151/5:20. Zone 2 is from 151/5:21 until 152/2:41. Zone 3 is from 152/2:42 until 153/0:02. Orders shown in bold have arrived at the buffer origin.

Notice that KAB/20 for S/O 126 has not arrived at the constraint and that is due to start prior to 151/5:20, the designated ending for zone 1. A quick look at the preceding part/operation's Inventory should help to locate the material and to expedite. If the material has not reached the constraint before the previous order has been completed, the next order on the list should be processed instead to prevent losing Throughput.

Managing the Buffer

What is needed now is a method for determining (1) whether current protection is adequate and (2) where to focus activities to improve. Figure 6.12 is a graphical representation of a buffer cross section frozen in time, with each zone representing 2 hours.

Each block represents an order and its timing in crossing the buffer origin. The numbers at the left represent minutes. The numbers at the bottom represent hours. The first part of order 100 was scheduled to cross the constraint at 6:00 and be completed at 6:17. Order 109 was to start at 9:00 and run until 9:15. Bold print indicates that an order has arrived at the buffer origin, while normal print indicates that an order has not arrived. Different zone profiles indicate different problems.

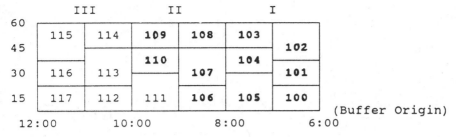

Figure 6.12 Graphical representation of the buffer.

Figure 6.13 indicates that all but two orders within the entire buffer have arrived, leading to the conclusion that the 6-hour buffer may be too long.

Figure 6.14 indicates that very few orders have arrived leading to the conclusion that the buffer needs to be more than 6 hours.

Figure 6.15 has a hole in the buffer where order 103 is missing, leading to the conclusion that some problem caused it to be late and that it needs to be expedited.

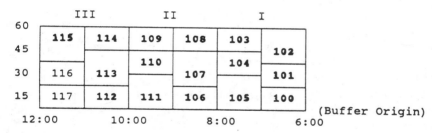

Figure 6.13 Buffer that is too long.

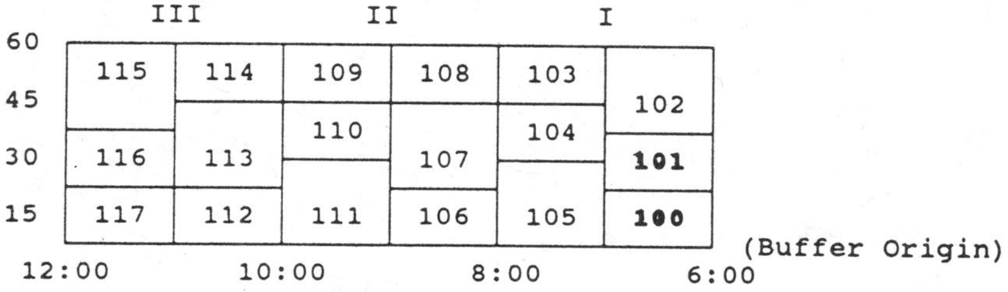

Figure 6.14 Buffer that is too short.

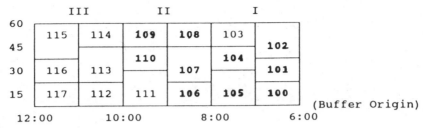

Figure 6.15 Holes in the buffer.

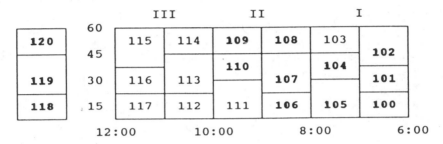

Figure 6.16 Material released too early.

The arrival of material that is not due within any of the zones it means that material release may be uncontrolled (Figure 6.16).

Individual buffer profiles can be reduced to percentages figures to indicate the health of the buffer profile. In Figure 6.17, after comparing the time that parts were supposed to arrive with the actual time of arrival, it is found that 90% of the orders have arrived by the beginning of zone 1 and that 75% of the orders have arrived before the beginning of zone 2. This means that 10% of the orders are being expedited.

Individual companies may find that this situation is acceptable. However, finding that zone 1 has a 60% rating may not be tolerable and is a signal that either the amount of protection allowed should increase, resulting in an earlier release of

```
Gating           III       II         I      Protected
Operation      -------|-------|-------   Resource
                          75%       90%
```

Figure 6.17 Buffer analysis by zone.

material and an increase in Inventory, or that whatever was causing the delays be eliminated or its impact reduced.

Measuring the Impact

Anytime an order has been expedited because of failing to arrive at the buffer origin by the beginning of zone 1, a note should be made as its location and to the cause of delay. These causes should be collated by location and a Pareto analysis done for those instances that impact Throughput the most. To understand the total impact on Throughput, these occurrences should be quantified in dollars and the lateness should be measured. The resulting formula is the dollar amount of the sales order to be impacted minus the raw material value times the number of days late, or Throughput dollar days (T$D):

```
Throughput × days late
```

Throughput dollar days are collected by location, by occurrence. Those resources and occurrences collecting the most T$D are the primary focus for improvements that should result in reducing Inventory while increasing the amount of protection available.

It is not always possible to fix the exact problem that has caused the delay. However, any method used to increase the amount of protective capacity available may offset the problem. For example, a specific rework problem is causing a delay in material getting to the constraining resource prior to zone 1. The rework is being caused by the lack of a fixture on a certain machine, replacement of which may take some time. However, unloading material to another machine that has excess capacity may release enough protective capacity to deal with this problem.

The Buffer Management Worksheet

Buffer management worksheets are organized to facilitate quantitative analysis and to ensure uniformity in data collection. Whenever a part fails to reach the buffer origin prior to the beginning of the expedite zone, it is expedited and the cause and location of the delay are recorded along with the amount of lateness incurred. This information is used in the Pareto analysis to determine what actions must be taken to reduce the amount of protection necessary and is an adaptation of the standard checksheet used to collect data and to quantify the type of defect.

The buffer origin designates the location of the buffer at a specific resource or at shipping. The buffer length gives the buffer manager an indication of the

Buffer Management Worksheet

Buffer Origin: ___Resource C_____   Date: ___10/25/91__

Buffer Length: _16hrs   Zone Distribution: I 75% II 90%

Name: _John Smith_____

Loc	Station	Cause	Zone I Arr.	Act. Arr.
R-1	123/30	Solder	154/2:15	154/5:15
R-1	123/20	Keyway	154/3:30	154/3:45
R-2	121/30	Out Mat.	154/5:10	154/7:10
R-3	126/40	Rework	155/1:30	155/2:30
R-1	131/20	Mach Dwn	155/2:30	155/3:30

buffer length used at a particular operation so that an idea of the amount of protection given can be established. The zone distribution gives the buffer manager an idea of the health of the system prior to the buffer origin.

The Buffer Management Reports

While the buffer management worksheet is used to collect data and can give a certain amount of insight as to the health of the system, some idea of the relative impact of an occurrence must be gauged by its impact on the bottom line. One can create a buffer management report by taking the input from the buffer management worksheet and determining the relative impact from the Throughput Inventory dollar days (T/I$D) equations.

Pareto Analysis

Pareto analysis is a specific type of histogram that gives an instantaneous picture of the priorities for improvement projects based on their history. The source of data is the buffer management worksheet, which is used to separate the "vital few" resources and causes, which create the majority of T/I$D, from the "trivial many." Pareto analysis can be accomplished two ways: the first is to discern which resources are collecting the most T$D and then to prioritize the problems within each resource that have caused the T/I$D to accumulate (Figure 6.18).

Obviously, the majority of T$D are being accumulated by resource R-2. A further Pareto analysis of the causes reveals that 55% of the delays were related to seals.

<u>Buffer</u> <u>Management</u> <u>Report</u>

Buffer Origin: ___R-5_____   Date: ___10/25/91__

Buffer Length: _16hrs   Zone Distribution: I <u>75%</u> II <u>90%</u>

Name: _John Smith_____

Loc	Station	Cause	Length	(T) Value	T$D
R-1	123/30	Solder	0.5 Days x	$1,000	= 500
R-1	131/20	Keyway	1.0 Day x	$2,000	= 2,000
R-2	121/30	Out Mat.	2.0 Days x	$1,500	= 3,000
R-3	126/40	Rework	1.0 Days x	$2,500	= 2,500
R-5	123/30	Rework	0.5 Days x	$1,000	= 500
R-1	131/20	Marking	1.0 Day x	$2,000	= 2,000
R-2	121/30	Rework	2.0 Days x	$1,000	= 2,000
R-5	126/40	Rework	1.0 Days x	$2,500	= 2,500
R-1	123/30	Rework	0.5 Days x	$1,000	= 500
R-1	131/20	Marking	1.0 Day x	$2,000	= 2,000

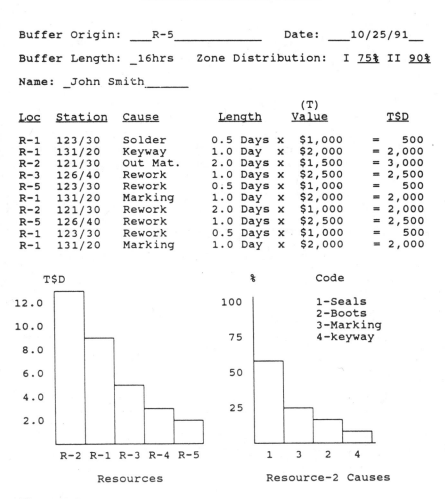

Figure 6.18 Pareto analysis by Throughput dollar days.

The Shop Floor Layout

Once the method of scheduling a factory for maximizing productivity has been found, some insight can be gained on optimizing the shape of the factory floor. While every manufacturing facility is different, some similarities definitely exist. The biggest change called for is usually in the area of materials location and in the provisions made to accommodate volume. In those areas directly in front of the constraint(s)—shipping and assembly operations where one leg is fed by the constraint—there will be a moderate amount of Inventory buildup. Inventory must be stored in such a way as to allow easy access for the implementation of buffer management. For those areas not located at the constraint, requirements for Inventory storage will be negligible. Material will be passing these locations very quickly and in small quantities (Figure 6.19).

Resource 6 has been identified as the constraint, so the buffer area has been established at this location. Buffers have also been created in front of shipping and in front of resource 9. This particular layout does not preclude the use of other techniques for increasing efficiency by lowering the amount of distance traveled, such as U-lines or group technology (GT) cells. Resource 6 may be a U-line, while resource 5 may be a GT cell. However, it must be remembered that converting to a process of higher efficiency at resources with excess capacity may actually result in a decline of profitability.

Rules for Non-Constraint Resource Utilization

Under the DBR process, the amount of Inventory at non-constraint resources will be very low. The method of determining priority is first in/first out (FIFO). Since the schedule for the constraint and the amount of protection required determine the release date, the sequencing of orders has already been determined and should be by arrival date/time to the non-constraints. Unless there is a problem causing an order to penetrate zone 1 in buffer management, this rule should not be changed.

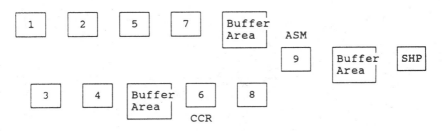

Figure 6.19 The shop floor layout. CCR = capacity-constrained resource.

Resources are activated when material is available for work, and deactivated when no material is available. Employees not having work in their area should spend time on employee involvement programs. However, they must be available immediately if work does arrive in their area. The work should be completed as fast as possible and sent to the next operation.

DBR in the Office Environment

Resources tend to interact the same regardless of the environment. An engineer can be considered a constraint, a near-constraint or a non-constraining resource. And while scheduling a factory and scheduling a group of engineers may seem worlds apart, from that perspective they are, in fact, one and the same. The same problems occur in much the same way and should be dealt with in the same manner. If the objective is to maximize productivity from engineering, then that resource (an engineer) who is limiting output must be exploited, and all other resources must be subordinated to the way in that it is decided to exploit the engineer's time. If an engineer who is in a critical position within the design process is slow or lacks the knowledge needed to complete designs quickly, this will present a problem in increasing the volume of the overall design process. At the same time, an engineer who is fast and knowledgeable but has excess capacity may be used to increase capacity at the constraint.

Problems in lead time reduction involve the same strategy as buffer management, in which those things that threaten the schedule for the constraint are identified by zone. While few companies have created route files and bills of material for the design process, most engineering projects include some indication of the amount of time required for completing each design, or group of designs, and a basic sequencing. There should be enough information to create a product flow diagram and a schedule. As in the factory, a physical constraint can be identified using current reality trees and by experimentation.

In an area such as accounts payable, little information is available for indicating the amount of time it should take to enter an invoice. However, most clerks can tell you exactly how long it takes. The sequencing of operations is usually very simple and well established.

Remember that implementing DBR in accounts payable or engineering just to improve the process may do nothing to increase profitability. There is, however, a growing issue over the shelf life of products and the speed at that designs are created and become obsolete. This is a strategic issue that must be addressed. The DBR process will help immensely in overcoming problems in reducing design times by isolating those problems that contribute immediately to the increase in design volume and to the reduction of design lead time. This is a key issue and must be addressed whether or not designs are immediately blocking the increase of a company's Throughput.

The DBR Advantage

Dr. Katsuhisa Ohno, one of the creators of the just-in-time process, has been quoted as not knowing *why* his process works, only that it *does*. The developers of the DBR system understood the impact of the dependent-variable environment and constraint management on the scheduling/improvement process and incorporated it to create a system that has been proven far superior to other programs, not just for scheduling the factory but for improving profits. While the impact of the dependent-variable environment on being able to produce a solution to the scheduling process and create a smooth product flow is important, the objective of the company is still to make money. The DBR process was designed to support this objective and therefore represents a tremendous advantage.

To summarize, the advantages include:

- Less inventory
- A more efficient and predictable improvement process
- Greater flexibility
- Less vulnerability to Murphy's Law
- A shorter implementation period
- Better manageability
- Higher Throughput
- Enhancement of the decision-making process
- Support and implementation of the five-step process

STUDY QUESTIONS

1. List and explain the disadvantages of Just-in-Time manufacturing.
2. Define the drum-buffer-rope (DBR) process and explain the advantages over more traditional scheduling processes.
3. What phases of the five-step process of continuous profit improvement are represented in DBR?
4. Explain what is meant by the term "placing orders on a time line," and how is it performed?
5. What two conditions must be met for placing orders on the time line?
6. Describe the buffer mechanism used in the scheduling process and how it is used.
7. Define the "rope" and describe its primary function.
8. What data is required to generate the data necessary to schedule the constraint?
9. How is the start time computed for the initial placement of orders on the time line?

10. What is meant by the term "leveling the load," and how is this ac-complished? Why must it be accomplished in this fashion?
11. What is meant by the term "rescheduling for time zero," and how is this accomplished?
12. How is the release date determined for orders that must cross the con-straint?
13. How is the release date determined for orders that do not cross the constraint?
14. Define and show graphic examples of the three different buffer types used in DBR. Explain how and why each is used?
15. Define and explain the difference between static and dynamic buffering.
16. What mechanism is used to focus efforts on reducing inventory and operating expense?
17. What is a buffer management report and how is it used?
18. What measurement is used for measuring the impact of those things that should have occurred but did not? How is it computed?
19. What measurement is used for measuring the impact of those things that occurred before they were scheduled to occur. How is it computed?
20. What is a buffer management worksheet and how is it used?
21. What are the rules for non-constraint resource utilization?
22. Explain the advantage of using DBR in the office environment.

7

The Attributes of a TOC-Based TQM Program

This chapter begins the process of developing a perspective on a TOC-based TQM program and identifying its attributes.

OBJECTIVES

- To introduce the principles and key components of the TOC-based TQM system
- To establish the TOC-based TQM strategy and requirements
- To define quality as a necessary condition and to understand why companies fail to meet their customers' quality demands

TOTAL QUALITY MANAGEMENT DEFINED

Traditional Total Quality Management (TQM) offers a management philosophy and structure designed to improve the profitability of companies through the practice of continuously improving all facets of each functional area, including the management process. The TQM umbrella includes the application of statistical methodologies that support a process of constantly monitoring the degree of variability within processes and for determining cause-and-effect relationships. Employee involvement as well as cross-functional management teams are used to manage the implementation of the improvement process and to focus on areas within the company that need improvement. Focusing mechanisms are designed to continuously improve the quality of products and activities while reducing costs.

ESTABLISHING THE NEED FOR CHANGE

American business must never lose sight of the fact that the goal of any business is to create wealth. But how do current-day TQM/TQC programs focus on this issue? Good quality is a requirement placed on each company by its customers. Unless a company is perceived by the customer as providing good quality at an acceptable price, the customer will probably not spend money for the company's products or services. However, fulfilling the requirement of good quality does not guarantee that the company will sell more product.

In order to create an improvement in profitability, sales must move up (increase), while Inventory and Operating Expense must move down (decrease). Once the requirement of good quality—even when quality is defined in its broadest terms—has been attained, any further increase in quality may increase only the *potential* for higher sales and not the sales revenue itself. How then must the requirement of good quality be satisfied while achieving the goal of increased profits? What actions must be taken to link the process of improvement to profitability?

If companies are to be successful at implementing TQM programs, they must make the connection between the measurements being improved on, such as reliability, complaint reduction and processing time and the impact on return on assets. If a company's major problem in dealing with profit is product quality—no matter how it is defined—if product quality goes up, then profitability should follow. As an example, if the quality of the products being produced at a given company is below the average market quality and has had a negative impact on sales, it would be expected that if quality were to be improved to a level equal to the average market, then sales would increase. However, continuing to increase quality that is already above that of the average market may not result in an increase in sales.

A regional director of maintenance for a major computer manufacturer was faced with a problem customer. The customer's perception was that the first time to repair (FTTR) and mean time between failures (MTBF) figures were unacceptable and that if they were not fixed, the account would be in jeopardy. Fortunately, the regional director had the presence of mind to gain the customer's perspective first and to determine whether it resembled reality. It was discovered that the customer's perceptions arose from problems that had occurred three years earlier. The problems had been resolved since then and yet the perception remained. If any money had been spent increasing the quality of the maintenance services offered, profitability would have moved down. Even worse would have been the impact on future sales through the loss of a customer. It must be understood that in order for profits to improve, companies must improve the right things.

To improve the right things, a company must focus on solving problems pertaining to profitability. And, for all companies, once a problem has been solved and an increase in profitability occurs, the same solutions will not continue to reap the same benefits indefinitely. Since every solution serves to invalidate itself over time, those indicators being improved today that result in an increase in profit may not result in an increase tomorrow. This can be quite a challenge.

Many companies, after going through a full TQM program, are disappointed at the resulting profitability picture. It may be that the measurements and focusing mechanisms used to determine what needs to be improved have not led them to a corresponding increase in profit. Many improvements may have resulted in increasing merely the *potential* of increasing profitability.

The key issue to be resolved is, can the traditional TQM methodology be improved? Can it be made to work faster, gain more profits or continue beyond current management strategies to solve problems for which solutions are not readily available? And, just as importantly, can the methodology be designed so that more companies can duplicate it?

. . .

To say that TQM cannot be improved breaks one of the basic laws upon which it was originally built: that anything can be improved.

. . .

THE TOC-BASED TQM OBJECTIVE

The objective of a TOC-based TQM is to establish an effective management system designed to implement the process of continuous profit improvement while meeting the necessary condition of good quality.

PRINCIPLES OF TOC/TQM

The principles of a TOC-based TQM serve as guidelines to help in understanding how to focus efforts in maximizing profitability through the implementation of TQM. A violation of these principles will result in undesirable effects.

Principle 1: Quality is a necessary condition.

The perception of quality for money spent is a condition that must be met before most people will buy a particular product. While an increase in quality will not always guarantee an increase in profitability, the condition of quality has a regulatory effect that, if not continually met, will definitely result in a decline of profitability.

Principle 2: Every solution will serve to invalidate itself over time.

Any solution that may have been valid at the time it was implemented will serve to invalidate itself once it has solved the particular problem it was designed to fix. If the problem is a flat tire and the solution is to mount a spare, once the spare tire has been mounted, the problem as well as the solution are no longer in effect. While this seems like a trivial matter, when applied to a chain of events that occur in a company, it means that once a problem has been solved, continued attempts to fix it would be utterly unproductive. Using the spare tire analogy, continually driving from store to store to buy a new and better tire to replace the one that went flat would be a complete waste of time. In the same way, continually massaging a cost accounting issue—even though the massaging has ceased to solve any problems—would be foolish.

Principle 3: The Throughput of the system is determined by its constraints.

As seen in the description of the process of continuous profit improvement, companies represent a chain of events. In any chain there is a weak link. It is the weak links that determine the limits to how much money can be generated.

Principle 4. The value of an activity is determined by the limitations of the system.

The limitations of the system ultimately determine how much money a company can make. Any activity must be weighed based on its impact on those limitations. Value is not determined by the frequency or the cost of the occurrence. An improvement in a resource that has excess capacity will not by itself result in an increase in the profitability of the company and may actually result in a *decrease* in profitability.

Principle 5. In a chain of events, the utilization of any resource may be determined by any other resource in a chain of events.

The interactions between resources and ultimately customer or forecasted demand determine the extent of utilization. However, since resources are used collectively to produce an overall effect of satisfying the customer, if one resource does not perform its specific function, it will ultimately block the creation of Throughput.

Principle 6. The level of Inventory and Operating Expense is determined by the attributes of the non-constraints.

Inventory and Operating Expense exist to create or protect Throughput. The characteristics of the non-constraints determine how much protection is actually required. A decrease in protection due to a lack of capacity at a non-constraint resource results in an increase in inventory, as jobs must be released earlier. Operating Expense goes up as overtime is needed to catch up. Material arriving late to the constraint does so because of a problem at a non-constraint, whether it be delivery of late material by a vendor or a quality problem at a product resource .

Principle 7. Resources are to be not merely activated, but utilized in the creation or protection of Throughput.

Activating resources to produce inventory that is not needed to create Throughput will ultimately result in an increase in inventory. An increase in Inventory will cause profits to decline. However, the major problem is the impact on Throughput. As Inventories go up, lead times are extended and Throughput declines.

REQUIREMENTS OF TOC-BASED TQM

The successful implementation of TQM requires

- Top management leadership and commitment
- Employee involvement and human resource excellence
- A Throughput orientation toward continuous profit improvement
- An orientation toward customer satisfaction

The key TOC-based TQM strategy involves team problem solving utilizing self-directed worker and management teams and valid/well-focused improvement processes and control mechanisms.

KEY COMPONENTS OF A TOC-BASED TQM PROGRAM

A TOC-based TQM program is a very broad system of management, which includes the following key components:

1. An orientation toward continuous profit improvement
2. A valid decision support mechanism
3. A customer-oriented quality focus
4. Local measurements that are in line with global measurements
5. A people oriented management system
6. A team approach to problem solving
7. A companywide focus on the five steps of improvement.
8. The religious use of current reality trees, evaporating clouds, future reality trees, prerequisite trees and transition trees in problem resolution at all levels
9. A program of supplier involvement and cooperation
10. An internal orientation toward customer–supplier relationships
11. A valid method of focusing improvement programs

12. The use of statistical as well as fail-safing methods for controlling processes
13. Long-term business focus and commitment
14. A prevention-oriented quality program
15. An unencumbered network of information exchange
16. A valid scheduling mechanism
17. A controlled program of variation reduction
18. A fully integrated system of profit improvement and quality management
19. An empowering approach to employee involvement
20. Tailor-made management and control strategies
21. Employee-based process ownership and commitment
22. A dynamic system for learning, managing and adapting to change
23. Comprehensive and well-focused education and training

These components are defined as follows:

1. An orientation toward continuous profit improvement

The objective of most companies is to make money. A healthy company is able to grow continuously. The TOC-based TQM system focuses on helping a company grow and on focusing efforts on continually increasing profits.

2. A valid decision support mechanism

Making valid decisions is the cornerstone to being able to manage in any company. As long as decisions are in line with the global goals and objectives of the company, the chances of success are far greater. Making valid decisions requires having access to valid data upon that to base the decision. The traditional decision support systems (via cost accounting) fall very much short of being able to accomplish this task. Information and data systems may need to be completely rethought. As an example, a valid system is needed to decide what product mix will maximize the profitability of the company. Traditionally, the solution has been based on the cost of goods sold, using various overhead allocation models. Emphasis was placed on understanding the "true cost" of a product and then subtracting that from the sales price. The product that had the largest margin was considered the most profitable, regardless of how much money could actually be made given the limitations provided by the environment. However, it has been found (see Chapter 5) that the information actually required is very much different from what has traditionally been used. Based on the Throughput generated per unit of the constraint, the decision support system for identifying optimal product mix requires knowing:

- The identification of the constraint
- The cost of raw material
- What price the product is being sold at

- How much time is used by the constraint in making the part

While most standard systems can provide details on the cost of raw materials, the sales price and the amount of constraint time used, few are capable of actually finding the system's constraints.

3. A customer-oriented quality focus

The first step toward meeting the necessary condition of good quality is to determine and set the quality policy. The customer buys the products and therefore determines what level of quality is required. The quality policy determines the quality level (the necessary condition) that is to be met. The quality function deployment program focuses the customer's desires on the processes and sets the standards to that resources are expected to perform. Once the quality policy has been set, visible and obvious management commitment from the top generates waves of commitment and quality attitude throughout the company. This, when supported by a clear and concise program of assigning quality objectives and the active elimination of the root causes for failing to meet customer expectation, will result in the necessary condition being fulfilled.

When a quality policy of meeting the necessary condition is not matched by a process of continuous profit improvement, the enthusiasm for maintaining the necessary condition will die. Management as well as employees will abandon those activities that do not lead to satisfying the profit requirement.

4. Local measurements that are in line with global measurements

Local measurements on that to build a valid decision system that will impact the global measurements correctly must be made for any system to produce the desired results.   The establishment of Throughput, Inventory and Operating Expense provides an intuitive as well as empirical basis for making local decisions and for focusing the improvement process.

5. A people-oriented management system

The TOC-based TQM process requires a people-oriented management system designed to instill self-ownership.   Authoritarian management styles do not effectively support a process of continuous profit or process improvement. The short-term benefits of fear-inducing or shaming tactics soon give way to long-term negative effects. People can be threatened with losing their jobs for only a short period of time before they begin responding in ways that are non-productive.

Management must be committed to solving problems, not assessing blame. Managers must learn to lead people either through group or individual activity by having them invent their own solutions so that the effects of the fear and "not invented here" emotions will be minimized. The TOC-based TQM management style is an open management process with clear and consistent objectives, well communicated and effectively explained to all. Emphasis is placed on a group-derived improvement process through open forums and the elimination of adversarial conditions and elitist management attitudes.

An effective TOC-based TQM management program will use effective combinations of motivational (humanistic) and scientific (quantitative) elements to effect control by "choice" rather than control by edict (force).

Trust is key. Employees and managers alike should be allowed to do their jobs without people looking over their shoulders. A committed management will be actively involved in helping employees do their jobs, but from a supporting role. The positive effects of having the president of a company visible on the shop floor just to listen to people's problems is well documented, and since employees tend to model the attitudes and behaviors of management, it is a great opportunity for an infusion of good will and TOC-based TQM world thinking into the entire corporation.

6.   A team approach to problem solving

It has often been said that the whole is greater than the sum of its parts and in managing complex organizations the most effective way to tap a company's human resources is through the use of self-directed worker and management teams where the generation of concepts and new ideas can be developed, fostered and implemented, and where manager and worker involvement/cooperation can be perpetuated. TOC-based TQM uses a team approach to bridge the interface between functional disciplines and eliminate the natural antagonism caused by different functional goals and work procedures. TOC-based TQM supports problem solving as well as long- and short-range planning through the rigorous use of problem-solving tools in group forums such as:

- The executive council
- The quality management council
- Cross-functional management teams
- Project teams
- Employee involvement teams

7.   A companywide focus on the five steps of improvement (see Chapter 2)

To produce a process of continuous profit improvement, a system of identifying the constraint and determining how to exploit it, subordinating the remaining resources to those activities designed to protect the constraint and then elevating and repeating the process must be understood and carried out by everyone in the company. Since every company operates as a chain of events, any function within the organization can block the process that improves the amount of Throughput being generated. Therefore everyone within the company must be exposed to the process and support it religiously.

8. The religious use of current reality trees, evaporating clouds, future reality trees, prerequisite trees and transition trees in problem resolution at all levels

Being able to create a process of continuous profit improvement means being able to find the constraint and then create and implement simple solutions that can be easily understood. Since 90% of a system's constraints will be policy oriented and not readily identifiable and since most constraints will be caused by erroneous or outdated thinking, it is essential that a system be implemented that organizes and focuses the thought process and then directs actions toward solving the core problem.

- *Current reality trees (CRT).* Used to identify core problems
- *Evaporating clouds (EC).* Used to break the assumptions causing the erroneous thinking that created the core problem
- *Future reality trees (FRT).* Used to predict the outcome of a specific change
- *Prerequisite trees (PT).* Used to uncover and solve intermediate obstacles to achieving the goal
- *Transition trees (TT).* Used to define those actions necessary to achieve the goal

9. A program of supplier involvement and cooperation

Every supplier's operation should be considered as an extension of the manufacturing facility and treated as such. Suppliers are valuable partners in business who can help tremendously in accomplishing the ultimate goal of implementing a process of continuous profit improvement. Relationships should be based on mutual interest and trust, not antagonism and distrust.

10. An internal orientation toward customer-supplier

Worker should think of themselves as both supplier and customer in the chain of events leading to the creation of Throughput, and act accordingly. As these relationships manifest themselves, worker attitudes and habits begin to change. The self-checking and next-process inspection procedures act to reinforce the quality of work from each resource by giving immediate feedback to the process responsible for creating the defect. Each resource (1) uses a short checklist to validate whether the work performed at the resource meets a certain level of quality before it is sent on and (2) has a separate checklist for validating that the work received from the previous resource meets certain quality levels. Lists are not extensive and add very little to the overall workload. Since most resources will be non-constraints, there will be no additional Operating Expense. This procedure offers the benefit of enhancing the smooth flow of material through the plant and creating additional visibility when one is trying to determine the cause of holes appearing in zone 1 of the buffer management system. It also ensures that a large number of people have viewed the quality of work performed before constraint time is wasted.

11. A valid method of focusing improvement programs

Variation reduction, setup reduction, total productive maintenance, engineering improvements and marketing programs all need ways of focusing to ensure that their results will have a positive and immediate impact on profitability. TOC-based TQM provides a valid method of focus, with profitability as the ultimate goal.

12. The use of statistical as well as fail-safing methods for controlling processes

Once an assignment of resource capability requirements has been made either by addressing customer's requirements via the quality function deployment system or the impact of demand on the relationship between resources, the degree of process control necessary can be determined. The regular use of statistical methods, such as statistical process control (SPC) and design of experiments (DOE), as well as fail-safing methods (Poka Yoke) to control processes will help to ensure that

- Resources are kept under control so that necessary conditions will be met.
- A maximum exploitation of the constraint or the effective subordination of non-constraint resources will occur.

If a process is under control, it is producing the best that it can produce without modification. Statistical methods help to determine when a process is out of control or due to go out of control in the future and give an indication of what may be wrong so that corrective action can be taken.

13. Long-term business focus and commitment

American business has long been conditioned to adopt a policy of focusing on the quarterly report to judge what actions are necessary to maintain a profit. For most businesses this is a fact of life. However, when the focus is on short-term cost reduction measures and not on concentrating on Throughput management, the results can be no less than catastrophic. How can a company fulfill the necessary condition of good quality and implement the five steps of improvement if it is narrowly focused on cost reduction? Because of a lack of a continuous profit improvement program, companies have had no choice but to be short sighted. But to sacrifice the future of the company to make the numbers look good today simply defies all logic. When a company that is successful at fulfilling the customer's requirements while implementing the five-step process, the quarterly focus problem goes away. It has become increasingly evident that it is possible to focus on short-term profits as well as long-term issues at the same time. But it must be done correctly.

14. A prevention-oriented quality program

Any attempt to fulfill the necessary condition of good quality must be oriented to defect prevention. It has been proven over and over that quality cannot be inspected into a product. To be effective, the focus must be on preventing mistakes from occurring in the first place. Products arriving at the constraint that have defects and use constraint time only to be scrapped will cause Throughput to decline immediately, and that amount of Throughput is not recoverable. Relying only on an inspection process in front of the constraint to catch defective parts would mean that an unacceptable number of defective parts would reach the constraint. A prevention-oriented quality program uses process control methods such as SPC and fail-safing to control the process. It also uses robust designs that are well within process capabilities.

15. An unencumbered network of information exchange

Blocking the flow of information will, by distorting reality, ultimately affect a company's ability to respond correctly. Making correct decisions and supporting a process of continuous profit improvement requires that the decision process be based on reality. Distorting reality means distorting the decision process, which will ultimately hurt the profitability of the company.

16. A valid scheduling mechanism

To be able to identify the limitations of the system now as well as in the future and therefore to support the decision process requires that the demand of each resource be known for the period of time for that the decision is to be made. To do this, a valid scheduling mechanism must be available that will outline the load, giving consideration for the limitations of the system and the relationship between resources.

17. A controlled program of variation reduction

Getting as close as possible to the target value is the objective of any SPC or TQM program. However, variation reduction for the sake of reducing variability is not by itself desirable. It should be part of an overall program of ensuring the current as well as future profitability of the company by fulfilling the necessary condition now, as well as in the future, and as part of the five-step process.

18. A fully integrated system of profit improvement and quality management

Every company needs a well-focused system for managing a program of continuous profit improvement and for fulfilling the necessary condition of good quality. The TOC-based TQM program provides a fully integrated system for accomplishing these objectives by insuring that:

- Every employee within the company knows what is expected and every activity is correctly focused.
- Valid measurement systems provide a solid basis for decisions.
- Everyone is able to participate in the process and understands what the indicators of success or failure are.

19. An empowering approach to employee involvement

It is the employees who determine the success or failure of any company and who hold the key to understanding what actions are really occurring. But often the chain of communication and the lack of ability to take actions that hamper any management strategy. Empowerment means giving employees the power to make decisions and take corrective actions where necessary to get the job done.

20. Tailor-made management and control strategies

Every company is different with respect to the problems encountered and the solutions required. TOC-based TQM recognizes that to be effective an innovative approach rather than a correlative approach of implementation is necessary. Implementing the same solutions as another company would miss the target altogether and result in the loss of money and not improvement.

21. Employee-based process ownership and commitment

Each process is identified and assigned an "owner" who is ultimately responsible for ensuring that his process is properly exploited if it is the constraint, or subordinated if it is a non-constraint, or that the quality characteristics assigned to it by the quality function deployment (QFD) program are met. Measurement systems are assigned emphasizing the commitment to global issues.

22. A dynamic system for learning, managing and adapting to change

Change, being inevitable and often illusive, requires a system of management that constantly shifts to meet changing needs. The TOC-based TQM system recognizes that any solution will become invalid over time and therefore must be capable of reacting to whatever change may occur. It must also recognize that while change is inevitable, people's resistance to change is also inevitable, and so a successful system must be prepared to prevent resistance.

23. Comprehensive and well-focused education and training

Education is a tool used not just to dispense information but to motivate and encourage people to invent their own solutions. When the learning process stops, companies begin to stagnate and ultimately fail to continue the processes that will help them the most. TOC-based TQM includes a lifelong commitment to training and education.

QUALITY AS A NECESSARY CONDITION

Poor quality is like inadequate cash flow. Whenever a company cannot find money to pay its debts, the decision process becomes skewed to the extent that the cash flow problem becomes a primary consideration in almost every decision. Even if the problem is solved, dealing with vendors who are hesitant to ship because of poor payment records can be just as disastrous. Companies that have found additional cash to feed the operation are still faced with poor credit. Vendors that have spent a large amount of time trying to get paid are hesitant to ship for fear of losing more money. If vendors will not ship, obviously customer orders cannot be filled—causing more cash flow problems. It is a vicious circle no company wants to face.

Like poor cash flow, the effects of a customers perception of poor quality will have an immediate impact on sales but it will also produce long-lasting residual effects. It may be more expensive and take more time to change a customer's perception than to actually increase product quality. Poor quality is a competitive edge issue that cannot be ignored. While good quality will not always increase the amount of money coming into a company, poor quality will definitely have a negative impact that can last for a long period of time. Product quality is a necessary condition and therefore should not be allowed to become the constraint to making more money.

MEETING THE NECESSARY CONDITION

The objective in meeting the necessary condition of good quality is to build and deliver products that meet customer expectations and requirements. To do this means being able to:

- Know what the customer requirements are and to convert them into product and process specifications
- Implement processes capable of meeting customer requirements
- Eliminate the blockers to meeting the necessary condition

The traditional causes for failure to fulfill the quality requirements of customers is quite large and includes the following:

- Preoccupation with short term profits and corporate mergers

- Using sales gimmicks as a substitute for solid quality, thereby creating customer perceptions that will not be met
- Not focusing on customer satisfaction
- Preoccupation with limiting customer complaints rather than focusing on customer satisfaction
- Lack of effective planning for long-term stability
- Failure to utilize the inherent abilities of workers to contribute to process improvement
- A culture permeated by attitudes of mistrust
- Massive bureaucracies with intricate checks and balances for controlling every action
- Focus on failure, with elaborate procedures for assessing punishment
- Management's inconsistency toward quality
- Elitism as a tool for management control
- The tendency to inspect defects out rather than building quality in
- Rigid systems that fight change
- Placing increased pressure on productivity without first correcting the problem
- Cutting expenses arbitrarily
- A dependence upon authoritarian management and single-track, single-minded, structures for control
- Antagonistic relationships with suppliers—price competition instead of cooperation—and large supplier bases
- Internal function–oriented, rather than customer-oriented, goals

While the symptoms and surface causes for failure to meet customer expectations can be very broad, the root causes are few. They include

- Conflicting goals and measurements
- Poor decision systems and support mechanisms
- Lack of understanding of how to meet the necessary conditions
- Lack of understanding of how resources interface and their resulting impact on the system
- The cost mentality
- Failure to understand how to motivate people

Of all the causes, probably the biggest problems to meeting the necessary conditions set by the customer and for meeting the profit objectives of companies stem from conflicting goals and measurements as well as poor or invalid decision systems and support mechanisms. Conflicting goals and measurements will result in organizations optimizing certain departments at the expense of others. As an example, the purchasing manager's effectiveness may be measured based on performance to standard cost. Management's objective is constantly to reduce the cost of raw material coming into the plant. The net result may be a large number of vendors, a high degree of variability in raw material and low-quality parts. This can lead to massive interruptions, an inability to successfully exploit the constraint or to properly subordinate the efforts of other resources and a reduction in overall profitability. But the purchasing manager gets a raise right before the plant closes.

The preoccupation with short-term profit objectives is usually a direct result of a failure to meet acceptable profit levels. Poor decision systems and support mechanisms will create a situation in which poor profit is a constant problem. The resulting turmoil will create constant shortages of qualified people to produce quality products as well as a morale problem, with people constantly looking over their shoulder for fear that they may be laid off next. Poor support mechanisms, such as the traditional method used for scheduling in the factory, will result in less than desirable results as well. At the first of the month, the pressure for delivery may be low. However, if a poor scheduling system results in a large backlog of late orders, meeting the necessary conditions of the customer may be in conflict with the obligations of the company in meeting investor and creditor requirements. High-quality parts are delivered at the first of the month, and low-quality parts are delivered at the end of the month.

A continuous focus on ways of reducing cost to meet profitability goals (the cost mentality) can play havoc on a company's ability to meet quality require- ments even when a company is not having a perceived profit problem. Employees are easily equated with cost outlets rather than opportunities for creating Throughput. This kind of attitude is quickly picked up by employees and trans- mitted to their work. Whenever cost is the primary issue, cutting corners becomes easier to justify. Required maintenance is overlooked or the need for new tooling is ignored.

Not understanding how resources interface or what the impact of a decision is from a global perspective can result in improvements being made in the wrong areas or in a total misunderstanding of how to schedule a production facility.

Without knowing what a customer needs, it is very difficult to fulfill the necessary condition. Failure to implement a program that details customer requirements, translates them into designs and product specifications and then into process specifications will ensure that the company's product-planning efforts will remain hit-or-miss. If quality is considered to be a necessary condition, then customer needs must be known.

The cost mentality is any learned response that serves to block the ability to look at reality in a common-sense fashion. It is the basis for most policy constraints and is probably the primary cause for not being able to implement a process of continuous profit improvement. It is the ultimate cause of conflicting goals and objectives as well as invalid decision systems.

CONTINUOUS PRODUCT IMPROVEMENT PROGRAMS

The degree of improvement of a product is a strategic issue, and whether or not a process of continuous improvement is carried out depends on its strategic implications. To ignore the improvement of a product in a highly competitive environment such as the computer industry is to court certain death for the product line or company. What characteristics a product must have to be salable one to five years from now cannot be ignored. Ignoring an opportunity to improve a product until the last minute simply because it is not the immediate constraint, or because the necessary condition is currently being met, while expecting to be competitive in the future is like sitting on a train track while a train is coming and being happy because you haven't yet been hit. Similarly, setting up a program of continuous improvement for a product whose life cycle is nearly spent or is expected to be stable for some time to come is probably a waste of time and money and will result in a decline of profitability. Where a company and its products must be in the future determines what improvement initiatives are implemented today. If it is expected that defect rates need to be lower in the future, then a program of constantly reducing variability may be necessary. If it is expected that products will need to be faster or smaller, then a process of constantly improving the product will be necessary so that necessary conditions can be met in the future. The quality strategy should include a decision to support or not support continuous product improvement and, if so, the degree of continuous improvement necessary.

PROCESS VARIABILITY

Process variation in manufacturing is inevitable. It is virtually impossible to create two objects that are exactly the same. It is the frequency and degree of variation that may have tremendous impact on the quality of products produced and the Throughput generated. Under traditional TQM, the objective of the statistical sciences is to track and support the reduction of the amount of variation that occurs in the system. Emphasis is placed on the magnitude and frequency of occurrence. The causes of abnormal variation for a manufacturing environment include *man* (personnel), *material*, *machine*, *method* and tooling. The causes of variation are further categorized into chance causes, which are considered normal, and assignable causes. Assignable causes create recognizable trends, are assumed to be correctable and is the subject of SPC.

Under TOC-based TQM, the degree of variability is not as important as its impact on Throughput, Inventory and Operating Expense. Variability is viewed as an issue in the five-step improvement process. If it causes a constraint or restricts the ability to properly exploit and subordinate, its relative importance must be elevated above the normal programs dedicated to the control and reduction of variability within processes.   Normal variability reduction programs should represent an on-going process designed to prevent quality from becoming the constraint or to ensure that the "necessary condition" is met as part of an overall strategic plan.

BREAKING THE NECESSARY CONDITION

In the event that quality has become the constraint, increasing quality may not be the immediate solution. Breaking the necessary condition may be more profitable in the short as well as the long run. In other words, if the playing field is set up for a given set of rules, either change the rules or find a game more to your liking using the resource capability you have. It is not necessary, nor often profitable, to assume that the rules cannot be changed. The distinction must be made at this point that what is being advocated is not running from a quality issue but looking for places to find more money without limitations in thought.

If a machine shop is unable to keep up with the technology being requested from current customers, it may elect to find a segmented market where the level of technology is not as stringent and sell part of its resource time into that market while making the changes necessary to get back into the high-technology market. Great care must be taken to ensure that the level of quality is competitive and acceptable to the new customer base. Otherwise, management may find that it now has a bad name in two markets.

THE TEAM CONCEPT

Few would argue that as the size and complexity of businesses grow, the ability to manage from a global perspective declines. While the effects may be devastating, the cause may simply be a decline in communication. Experience has shone that in the traditional company, the larger an organization, the more isolated its members. People become discouraged as communication drops off because they are unable to cope with the magnitude of the problems and because they do not feel a part of the process. Individual departments develop a myopic view of the world and are unable to function to maximize the whole of the organization. Managers as well as supervisors are forced to rely more and more on invalid localized measurements and decision mechanisms that can have a devastating impact on profitability.

While large companies may resemble lumbering dinosaurs, small companies often represent cohesive entities. They become successful because management is able to readily view individual activities/decision processes and their impact on profitability from a global perspective. Communication is greatly enhanced by the smallness of the organizational structure, where technology and information is easily shared.

Managing large companies requires a method for increasing communication and individual worker input while focusing efforts to maximize profitability. Work teams revitalize the small company perspective by increasing communication and creating the opportunity to manage globally. These consist of 6 to 15 highly trained individuals with shared responsibility for a finished work segment. Each individual contributes to the whole by bringing his or her unique ability and by creating an entity that is greater than the sum of its parts.

Work teams take many forms including:

- Cross-functional management teams
- Employee involvement teams
- The executive council
- Steering committees
- Self-directed work teams

It is important to ensure that each individual is able to contribute in a way that is constructive to the entire team. Each team member is trained not just in working with teams and team concepts such as brainstorming and the TOC thinking process; time is allotted to ensure that each member is technically qualified to contribute as well. Engineers are trained to be better engineers, while quality, production and purchasing people are trained to understand more about their functions as well. Each must function with a valid decision-making process and focusing mechanism.

How a specific team is used depends on what function it is asked to perform. A strategic decision to support a process of continuous product improvement will dictate that a different improvement process be used than for a team whose function is to increase profitability. The benefits of using the team concept are increased communication, a ready access to highly qualified people, a better sense of self-determination among the team members and a higher probability of problem resolution.

STUDY QUESTIONS

1. What is the objective of TOC-based TQM and how is it accomplished?
2. List and define the seven TOC/TQM principles.
3. Why is it that every solution will inevitably invalidate itself over time?
4. Define quality as a necessary condition.
5. List five key TOC/TQM components that are directly related to TOC.

6. What are the prerequisites to meeting the necessary condition?
7. List and explain three root causes for why companies fail to meet the necessary condition of good quality.
8. Define the term *cost mentality*.
9. Under what circumstances is the impact of variability magnified?
10. What advantages does the team concept provide to the implementation of TOC?
11. List and define five types of work teams.

8
Developing a TOC-Based
TQM Infrastructure

In this chapter an infrastructure for the implementation and establishment of the improvement process is created.

OBJECTIVES

- To create the organizational/communications structure and assign responsibility
- To establish insight into the documentation strategy for formalizing the process
- To introduce the concept of feedback and control mechanisms in completing the communications structure and to understand how they are to be used under the TOC-based TQM strategy

THE MANAGEMENT SYSTEM IMPERATIVES

To be effective, any management system must have:

- An effective means of relaying information
- A valid decision process
- A valid set of measurements
- An effective feedback mechanism
- A method for identifying, prioritizing and solving problems

Until now, the text has been devoted to changing the paradigms associated with traditional Total Quality Management that deal with the decision process, measurement systems and methods for identifying, prioritizing and solving problems. What must now be addressed are methods of developing and maintaining communication such as the organizational structure, documentation and traditional feedback mechanisms, which at may or may not require enhancement to support a global perspective.   Each of these issues must be addressed within the context of the five-step improvement process.

THE TOC-BASED TQM ORGANIZATIONAL STRUCTURE

The first task in creating the TQM II management system is to develop the organizational structure designed to support the system and to assign responsibilities.   The organizational structure should promote effective communication between organizations and ensure proper and timely feedback when problems occur. This will include:

- The TOC/TQM executive council
- The quality management council
- The cross-functional management team
- The project team
- The employee involvement and quality circle teams

The TOC/TQM Executive Council

The TOC/TQM executive council consists of the top executives from the company, including the president and representatives from the major functions of finance and accounting, manufacturing, engineering, quality, sales and marketing, distribution, and customer service.   It is important for every function to be represented.   Their responsibility is to formulate corporate policy and to oversee the program and its implementation.   From an executive level, this group determines the direction the corporation will take in implementing the five-step improvement process.   This group actively participates in group sessions and routinely uses current reality trees, evaporating clouds, future reality trees, prerequisite trees and transition trees in the development of the action plan.   A group forum is important.   While every participant should religiously prepare its own documentation, a group consensus can be developed by using these tools in open sessions while developing the corporate plan.   The head of the TOC/TQM executive council and primary facilitator should be the president of the corporation.   This group determines the strategic directions the corporation will take and the way in that money will be spent.   Elevating a constraint may necessitate buying a new machine worth millions of dollars or changing a major corporate policy.   It is of prime importance that the chief executive for the corporation chair these forums.

The Quality Management Council

The quality management council is established to ensure that by setting direction and establishing policy the quality of products and services being offered to customers are competitive. It has the task of developing, assigning and monitoring major quality improvement projects and for providing support where necessary to maintain a process of ongoing product or process improvement when necessary. The objective of the council is to prevent quality issues from becoming the constraint to making money. The quality management council is responsible for the quality function deployment program, among a number of different programs, and should be concerned about product safety, reliability, availability, maintainability and conformance to specification as well as customer perception. The quality management council develops, assigns and monitors quality improvement projects, works to identify and plan training requirements, sets measurement criteria, monitors the rewards management system and establishes and monitors the quality circles program.

Cross-Functional Management Teams

The cross-functional management team consists of managers from each function and has the responsibility to ensure that the action plan set by the TOC/TQM executive council and the quality management council is carried out. Activities are assigned to ensure that the constraint is properly exploited and that all other activities are subordinated to it. The cross-functional manage-ment team offers quick access to all functions for faster action when problems occur and, when the constraint is elevated, it reduces the probability that inertia will cause problems by exposing a larger number of people and viewpoints to these same problems. The cross-functional management team also uses current reality trees, evaporating clouds, future reality trees, prerequisite trees and transition trees in individual as well as group forums for the development of action plans and in solving problems. Note that in small companies the TOC/TQM executive council, quality manage-ment council and cross-functional management teams may be the same people.

The product design review team is a form of cross-functional management team and may include design, quality and manufacturing engineers as well as members from production, marketing and purchasing. Products are reviewed in a phased approach to ensure that all functions have input to product designs.

Project Teams

The occasion may arise when a project team is necessary on a temporary or on-going basis to develop methods of exploitation, such as for setup reduction or engineering changes; to study problems involved in subordination such as a quality issue; or to insure that the necessary conditions are being met. The project team is used to place a multitude of resources and technologies to work within the same group in solving short-term problems. The project team should also be aware of how to implement current reality trees, evaporating clouds, future reality trees, prerequisite trees and transition trees so that solutions will be focused and outcomes predictable. The structure of the project team is horizontal in nature, with membership being granted through appointment by the members of the cross-functional management team.

Employee Involvement (EI) Teams and Quality Circles (QC)

Employee involvement teams and quality circle groups include 3–15 shop, technical and supervisory people who meet on a regular basis to discuss problems involving exploitation and subordination as well as the requirements of the quality function deployment (QFD) program. Employees are taught the use of the elementary tools to problem solving such as statistical process control, Pareto analysis, current reality trees, evaporating clouds, future reality trees, prerequisite trees and transition trees. Additional tools may include fishbone charts, scatter diagrams and histograms. They are also taught to understand the impact of their actions in the dependent-variable environment and the use of the buffer management and the control systems associated with it. EI teams as well as QC circles may be headed by a team leader, who is usually the foreman or line supervisor acting as a group facilitator.

The effectiveness of the employee involvement effort in supporting the continuous profit improvement program and the extent of its usefulness will depend on where the primary constraint is and whether a strategic decision has been made to support a continuous product or process improvement program. If the constraint is in the market, improving production will not improve profitability unless it is a competitive-edge issue that is creating the market constraint and that can be overcome through a more effective use of production. As an example, if the core cause of a market constraint is identified as a long lead time caused by production, then using the IE team to solve the lead time problem should prove an effective use of its time. If a strategic plan has been set that will require a continuous increase in quality and it can be helped through the continuous reduction of variability, then the QC group would be useful in supporting this effort as well. If a strategic decision is made to enter a new market segment utilizing non-constraint resources but is being blocked by a resource that is being driven to constraint levels by the proposed new forecast, then the project team or EI team could be useful in reducing the amount of load on the newly constrained resource. In facing another strategic issue, a decision to bring the constraint inside the plant must be met with a plan that considers where the new constraint will be located and what kind of support will be required from the non-constraints. The

EI team and project teams are also useful during the initial implementation of TOC-based TQM.

One of the key issues of the team approach is to maximize creativity and increase communication. While the EI teams and QC groups may not always be able to contribute to the overall profitability of the company and may not be as active under TOC-based TQM, the effects of communication and peer pressure should reinforce the TOC-based TQM thinking process.

ESTABLISHING FUNCTIONAL RESPONSIBILITIES

Sales and Marketing

Sales and marketing participate in the implementation of the TOC-based TQM program by:

- Determining and selling the right product mix to maximize income.
- Helping to establish multiple market segments to safeguard resources.
- Helping to set product prices using valid methods.
- Determining whether products fit customer requirements through the analysis of feedback.
- Determining what quality characteristics will be needed to fulfill the necessary condition of good quality in the future.
- Determining the relative priority of product characteristics.
- Helping in the overall implementation of the five-step improvement process.
- Selling internal constraint capacity.
- Continuously examining and updating the sales and marketing policy book to eliminate policy constraints.

Manufacturing/Production

Production participates in the implementation of the TOC-based TQM process by:

- Maximizing constraint utilization through the creation of valid schedules and by supporting setup reduction, maintenance and SPC activities.
- Managing the buffer management and work improvement programs.
- Participating in design review to maximize constraint utilization and to ensure proper subordination of non-constraint activities by identify-ing critical issues such as the clearness of design specifications, equipment requirements and hard-to-obtain materials.
- Participating in the development of process specifications.

- Supporting worker and process certification functions.
- Producing to specification and schedule.
- Helping in the overall implementation of the five-step improvement process.
- Developing and executing the manufacturing/production management function.
- Continuously monitoring and updating the manufacturing/production policy manual to eliminate policy constraints.

Engineering

Engineering participates in the implementation of the TOC-based TQM process by:

- Developing products that meet current and future requirements for reliability, maintainability, performance, produceability and test-ability.
- Supporting project team activities for improving constraint utilization and for maintaining proper subordination of non-constraint activities.
- Participating in the supplier certification program.
- Participating in relevant market segmentation programs.
- Participating in design review activities.
- Helping in the overall implementation of the five-step improvement process.
- Developing and executing the engineering management function.
- Continuously monitoring and updating the engineering policy manual to eliminate policy constraints.

Finance and Accounting

Finance and accounting participate in the implementation of the TOC-based TQM process by:

- Participating in the supplier certification program.
- Administering the Throughput justification program.
- Helping in the overall implementation of the five-step improvement process
- Developing and executing the finance and accounting management function.
- Continuously monitoring and updating the finance and accounting policy manual to eliminate policy constraints.

Customer Service

Customer service participates in the implementation of the TOC-based TQM process by:

- Helping to establish multiple market segments to safeguard resources.
- Helping to determine whether products fit customer requirements through the analysis of feedback information.
- Helping to determine what quality characteristics will be needed to fulfill the necessary condition of good quality in the future.
- Helping to determine the relative priority of product characteristics.
- Helping in the overall implementation of the five-step improvement process
- Participating in the design review process
- Developing and executing the customer service management function.
- Continuously examining and updating the customer service policy book to eliminate policy constraints.

Quality Assurance

Quality assurance participates in the implementation of the TOC-based TQM process by:

- Offering training and consulting services for insuring the availability of qualified employees and for maintaining trouble-free processes.
- Participating in relevant customer relations programs.
- Administering the audit and survey programs.
- Managing the metrology program.
- Administering the inspection and test programs.
- Participating in the design review process to ensure proper subordina-tion of the quality activity.
- Participating in the supplier certification program.
- Helping to set and administer the continuous product improvement strategy.
- Establishing and maintaining the process capability program.
- Establishing and maintaining the statistical process control program.
- Helping in the overall implementation of the five-step improvement process.
- Developing and executing the quality assurance management func-tion.
- Continuously examining and updating the quality assurance policy book to eliminate policy constraints.

DOCUMENTATION

The next task in creating the TOC-based TQM management system is to ensure that a well-established system of documentation exists that is designed to support those activities required to fulfill the necessary condition and the implementation of the five-step process for all who are involved. Specific documentation includes:

- The corporate policy manual
- The quality assurance manual
- The functional procedures manual
- Process specifications manual

The Corporate Policy Manual

The corporate policy manual defines the corporate goals and objectives and establishes guidelines for the development of corporate procedures and strategies. It outlines the importance of quality as it relates to the customer and defines the method for insuring that the customer's quality objectives are met. It also sets the stage for focusing on Throughput and managing through the five steps of continuous profit improvement.

Typically, it includes guidelines for:

- Marketing and the development of market segmentation strategies or methods for determining product mix.
- Engineering and methods of new product introduction.
- The role of quality assurance.
- Establishing methods of process planning and control.
- The implementation of the measurement and the decision processes.

An example of a guideline is requiring the use of a phased approach for new product introduction that takes into consideration quality function deployment issues as well as the dependent variable environment or a statement requiring the use of Throughput, Inventory and Operating Expense as the basic measurement system. Special interest is given during the implementation process to changing the corporate policy manual to include the new methods of focus and operation.

Typical policy statements include:

- Quality is to be viewed as a necessary condition and is to be supported by the QFD process

- Establishing a program to support continuous profit improvement will be the rule and will be pursued by all functions
- Throughput, Inventory and Operating Expense are the measurements of choice and will be adopted for use by all functions and incorporated into the decision process
- The necessary level of quality is to be built in and not inspected into the product
- TOC-based TQM will be a key consideration in source selection, and vendors will be treated as partners
- TOC-based TQM concepts and practices are to be ingrained throughout the company through tailored, continuous, lifelong training for everyone, starting with top management
- Dedicated, competent and involved employees will be recognized and rewarded appropriately

The Quality Assurance Manual

The quality assurance manual documents the quality plan to include procedures for:

- Inspection and testing to establish the relationship between the products being produced and the QFD requirements, and to support the sub-ordination process.
- Audits and surveys to give feedback to managers of the overall effectiveness of the management systems, the degree to that they follow corporate guidelines and whether they are being followed.
- The metrology program to develop and regulate testing devices.
- Non-conforming material disposition to determine whether products should be scrapped or reworked.
- The overall statistical methods program to include the type of methods to be used and where and how they are to be focused.
- New product planning to ensure that the customer's needs are fulfilled and that the limitations and opportunities provided by the environment are not overlooked.
- Supplier relations to ensure that products and services provided by vendors meet with established guidelines for product quality.
- The subordination of the quality process.
- Establishing feedback mechanisms such as next-process inspection and fail-safing devices.
- Determining where and how to focus variance reduction programs.

The Procedures Manual

The procedures manual describes the responsibilities and method of operation for each functional organization within the company. Specific interest is given to insuring that each function has a well-documented method of determining how to best exploit or subordinate to the constraint and to determine what actions are required to meet the necessary condition of good quality.

The procedures manual documents the decision processes to be used and measurement systems supported. Included are instructions for:

- The implementation of the maintenance, setup reduction and variation control programs.
- The day-to-day operation of materials management, sales and marketing, engineering, and finance and accounting.
- Methods of scheduling and controlling the shop floor.
- Instructions for managing employees and conducting employee involvement programs.

The Process Specifications Manual

The process specifications manual is the results of converting product specifications into process specifications within the QFD program and is the mechanism by that the process requirements are relayed to individuals within each process. It contains detailed instructions and tolerance information so that workers can accomplish the task of production. Included in the manual are instructions for assembly, fabrication and machining as well as the acceptable limitations. Typical process specification manuals for a printed circuit board manufacturer might contain pictures of how an acceptable or unacceptable solder joint looks. In preparing the process specifications, the preparer should avoid presenting vague information requiring unnecessary interpretation by the operator. He or she should cover thoroughly the important product characteristics and explain why a certain process is to be completed in a certain manner or why a certain specification has been set.

Feedback and Control Mechanisms

Feedback and control mechanisms are designed to relay information about a system's performance from a local as well as a global perspective and to offer possible alternatives or enforce correction. It is at the heart of the basic TQM program and holds a similar position within TOC-based TQM, and it offers a focusing mechanism for determining where and how to implement the old methods such as "Murphy-proofing," self-inspection, next-process inspection and statistical methods but also has some unique methods for setting priorities and determining global impact.

The Concept of Source Inspection*

Source inspection is an important ingredient in the overall TQM process. Conceptually, the emphasis is on inspecting and correcting problems as close to the object process as possible. Operator feedback is immediate, inspection levels reach 100% without additional operating expense, and social pressures require performance. Note that while the general trend is to move away from source inspection by designing better processes and products, the need for source inspection will persist until it can be eliminated.

Fail-safing devices (categorized by purpose and method) prevent defects from occurring: control devices prevent defects; warning devices give notice of pending defects. A machine modified to allow a part to be installed only one way is a control device; a buzzer that sounds when a defect is about to occur is a warning device. Fail-safing devices include pressure-sensitive switches, thermometers, electrical current sensors, light-activated switches, and counters.

Self-inspection utilizes checklists for operators to ensure that certain critical operations have been done correctly. Operators inspect their own work after the process is completed.

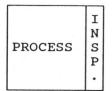

Next-process inspection uses the next person in the process to inspect the work of the previous process. Discovered defects result in the work being returned immediately to the causing departments or operations.

*Adapted with permission from *Production and Inventory Management Journal, 32*(1):7 (1991).

Statistical Process Control

In statistical process control (SPC) the emphasis is on continuous process monitoring; the assumption is that a process under control will yield acceptable product. The operator establishes the average value and range of deviations from the mean in a process and plots those values in the form of X-bar and R charts (Figure 8.1).

Precontrol limits are set and sample measurements are taken from the products being produced. The operator then monitors the outcome of each plot or group of plots to determine if a trend is developing or if the process is out of control. If it is out of control, the process is stopped, the cause determined and a solution implemented. (See Chapter 10, "Statistical Process Control.")

Precontrol, a simplified version of SPC, focuses more on process control/ capability verification and less on charting (Figure 8.2).

Precontrol process lines (PC lines) are set in the middle half of the product specification width. Process capabilities are established at the beginning of production by measuring five consecutive occurrences, each of that must appear in the green zone. Periodic sampling indicates whether the process is under

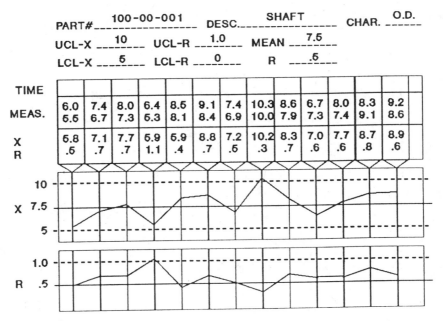

Figure 8.1 The X-bar and R charts.

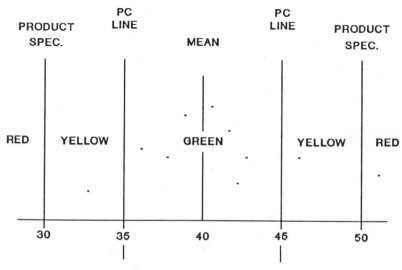

Figure 8.2 Precontrol.

control. (The frequency of sampling is determined by dividing the time interval between failures by 6.) Two units are measured during each sample, and the process is stopped if two measurements are in the yellow zone or one is in the red.

Since source inspections provide immediate feedback, these types of mechanisms are excellent for monitoring the most critical portions of the factory which need to react quickly to problems that impact Throughput and to include:

- Primary and secondary constraint(s)
- Resources that feed the constraint directly
- The chain of resources leading from the constraint to the sales order
- Those resources designated as critical to the QFD program

Buffer Management

Buffer management provides information on the success of the overall program from a global perspective by aggregating the impact of problems that will affect the buffer origin and provides a roadmap for corrections, as seen earlier. Also seen earlier, the implementation of Throughput dollar days and Inventory dollar days as control measurements forces action from offending resources to make corrections.

Audits and Surveys

Audits and surveys were originally designed to provide an indication to management of the effectiveness of the overall system. They concentrated on whether procedures had been put into place that would gain the proper result and whether or not these procedures were being followed. Key issues included laws and regulations, employee and customer safety, and conformance to specification. Of prime importance was determining whether or not proper decision mechanisms and measurements were being used.

There is a warning that must be given with regard to the effectiveness of the audit/survey program. Audits and surveys look at a broad spectrum and tend to widen the overall focus of management. Since very few changes are needed to support the process of continuous profit improvement, attempts to implement the many irrelevant changes prompted by audits and surveys may result in the unnecessary spending of money, and profitability may actually decline. The focused approach of using current reality trees, evaporating clouds, future reality trees, prerequisite trees and transition trees identifies what needs to be changed and how to accomplish it in far less time. Audits and surveys should be used sparingly to gain an overall view as to the validity of decision processes and procedures being used and whether they are being followed. The objectives should be to get an indication of what problems may surface in the future.

Inspection and Testing

The inspection and testing process refers to the process of determining whether a product's characteristics conform to specification, and then making a disposition and recording the data. It is usually performed by a group outside of the production organization. Under TOC-based TQM, the basic concept of inspection and test has not changed from traditional TQM. Inspection and testing is still the method of last resort because of the length of the delay in feedback to the offending operation. However, there are still times when a more detailed procedure using inspection and testing is needed by qualified people to ensure product compliance.

Like source inspection, the method of focusing the inspection and testing process has changed to comply with the QFD program and the realities dictated by the dependent-variable environment.

Benchmarking

Benchmarking has been described as making a comparison of how well one corporation is performing in relation to another. Similar operations are compared to determine whether improvement is available or should be warranted and to determine what, if anything, can be done to improve. The usual benchmarking endeavor includes comparing cycle times or reject rates for similar operations.

There is danger in this process, too, because the approach used is correlative. Since most corporations differ as to where the constraint is, a particular comparison might prove useless. Unless both companies have the constraint located in the same resource and are determined to exploit and subordinate in the same fashion, there may be no similarities to compare. Each company would be attempting to maximize different processes.

The net effect would probably be that improvements would be made in the wrong places. If money was spent, profitability would decline.

Companies seeking to benchmark should be very specific in what they are looking for. The objective is to maximize profitability, not improve every resource. It might be interesting to see how other companies have exploited similar constraint or near-constraint resources, to look at methods of subordination and to see how product presence in the market might compare among competitors.

Process Certification Program

Process certification is designed to ensure that a process is capable of consistently producing to specification and that each resource understands its role in the exploitation and subordination process. The certification of a process includes ensuring that:

- The proper documentation is available
- People have been properly trained and are performing to process specification and documentation
- Test equipment and tools are available and the calibration program is in place
- Workmanship standards are released
- Inspection criteria have been determined and inspectors prepared
- Process capability studies have been completed and the process is found capable
- Process controls are in place and functioning
- Employees are capable of making valid decisions in support of exploitation and subordination

Workers who do not understand the impact of their decisions or are unable to make decisions from a global perspective will not be able to perform well.

SUMMARY

The TOC-based TQM infrastructure has a significant impact on the overall implementation of TOC-based TQM by creating an organizational structure and acting to formalize the process. It enhances communication at a management as well as an operational level and ensures that action will be taken to solve problems. It serves to immediately indoctrinate new employees into the process and provides adequate documentation for them to learn what is expected of them and how they fit into the overall program.

STUDY QUESTIONS

1. List and explain the five characteristics of a good management system.
2. List at least five responsibilities of the following organizations: Sales and Marketing, Manufacturing/Production, Engineering, Finance and Accounting, Customer Service, Quality Assurance.
3. What are the two primary objectives of the TOC-based TQM documentation system?
4. What is the objective of the corporate policy manual, and what kinds of statements would one expect to find?
5. What system of documentation is recommended under the TOC-based concept, and what are the objective and contents of each?
6. List and define six forms of feedback mechanisms used under TQM.
7. What warnings should one heed regarding the use of the audit and survey programs?
8. What problems can be generated through implementing a benchmarking program?
9. When is a process considered to be certified?

9

TOC and Product/Process Design

This chapter introduces key issues associated with the design process and establishes a platform for designing products that will meet the necessary conditions of good quality while at the same time will supporting the process of continuous profit improvement.

OBJECTIVES

- To create the process for converting customer requirements into product/process designs so that the necessary conditions of good quality can be met
- To ensure that the design process properly supports the exploitation, subordination and elevation phases of the process of continuous profit improvement

DEFINING THE DESIGN FUNCTION

The function of product/process design is to convert customer needs into product/process specifications. How well a company can accomplish this task, in many cases, will determine whether or not it will continue to meet the necessary condition of good quality and remain a profitable entity. Those companies failing to continually meet the requirements of the customer soon become extinct. It is necessary to understand what the customer wants or needs and then be able to convert these wants and desires into instructions for those who must create the product or perform the service. It is desirable to accomplish this in a way that will maximize the profitability of the company.

QUALITY FUNCTION DEPLOYMENT (QFD)

Product quality is defined as delivering what the customer wants, on time and free of defects. To be effective, a quality program should be able to recognize when a product is perceived as no longer adequate in the customer's eyes and predict when it will be inadequate in the future, so that corrections can be made. All efforts should be made to ensure that the customer's perception of the quality of goods and services received is at least equal to that of the competition to prevent poor quality from becoming the constraint to making more money. To do this, one must know the customer's current and future desires and perceptions as well as the degree to that the competition is currently satisfying them and an estimate of how well the competition will satisfy them in the future. Once an understanding of the customer's needs has been developed, in order to meet those requirements, each function must understand what part it plays in fulfilling the customers needs.

The quality function deployment (QFD) program creates an organizational structure and control method for managing the coordinated development of products and services based on customer demand. It assigns product or process characteristics that are desirable to meet the demand and identifies those functions/organizations and activities necessary. Emphasis is placed on the use of the seven new management planning tools to obtain better product definition, communication and documentation during the design process so that design/redesign is held to a minimum. In addition, the QFD program compares those requirements to how well the competition is meeting them.

PLANNING FOR TOTAL QUALITY MANAGEMENT (THE SEVEN NEW TOOLS)

One of the problems inherent in the Shewhart planning cycle of plan-do-check-action is trying to determine what to plan. The management planning tool concept uses an intricate system of tools that are designed to create a detailed understanding of complex problems and to capitalize on the creative abilities of managers and employees in investigating and solving them so that effective plans can be created and executed. It begins with an assimilation of data into correlative and logical structures, and ends with a process for scheduling and controlling the implementation plan.

These tools include the

- Affinity diagram
- Relations diagram
- Tree diagram

- Matrix diagram
- Matrix data analysis chart
- Process decision program chart
- Arrow diagram

Each tool fulfills a specific role in the overall planning process and is used in a group forum to facilitate a group-accepted solution and a wider understanding of the issues involved.

Affinity Diagram

The affinity diagram is an intuitive process for developing ideas and examining complex issues that uses brainstorming and cards to organize input from various sources. To create the affinity diagram (Figure 9.1), the following steps should be used:

1. Select a group of people with a common interests and familiarity about the subject to be examined. Assemble them and give each member a stack of cards.
2. State the issue to be examined and give the group 10 minutes to respond by writing one idea on each card. Members should avoid communication.
3. After 10 minutes, each member in succession reads one idea and places it on the table. There should be no discussion of ideas at this time. New

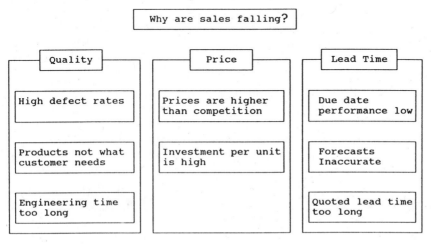

Figure 9.1 The affinity diagram.

ideas generated during this process should also be written down and
presented.
4. After all ideas have been generated, the cards should be arranged in
    related stacks and a common label selected for each stack.

The objective of the affinity diagram is to maximize the number of ideas
being generated and then to categorize them for further examination.

Relations Diagram

When studying complex issues, it is often necessary to look for relationships and
patterns to begin to develop so that they can be better understood. The relationship
diagram is a logical method for exploring relationships between factors. It is commonly, though not necessarily, used in conjunction with those ideas and concepts
developed in the affinity diagramming process. A central idea, concept or problem
is presented, and then all the logical connections to these issues are de-termined.
The logical connections can sometimes go through numerous levels before a
complete picture of all the relationships involved is discovered. Each level may be
a prerequisite for determining the relationship of the next. The two major uses of
this method include multiple- and single-level problem solving and can be
effective in addressing quality, management policies, production issues and design
problems. To create the relationship diagram, the following steps are followed:

1. Select and assemble a group of people with a common interests in and
    familiarity with the subject to be examined. This will probably be the
    same group used for the affinity diagramming process. (It may be
    appropriate for the group facilitator to have attempted creating the
    diagram on the same subject, independent of the group, so that process
    time can be minimized).
2. Select and define the central issue to be examined and write this on a card
    or on a blackboard or dry-erase board to be displayed in a central lo-
    cation.
3. Brainstorm for ideas on cause or prerequisite relationships, writing each
    concept or idea on the board in the proximity of the central issue. Add
    arrows to build a picture of the relationships of ideas. Repeat the process
    for new ideas as they are developed to create a broad picture of all
    relationships and to look for major issues and root causes to be examined
    more closely.

Figure 9.2 illustrates a completed relationship diagram in which the objective
of the exercise is to determine the relationship of the batching, material and
resource utilization policies for the competitive-edge issues of quality, lead time
and price.

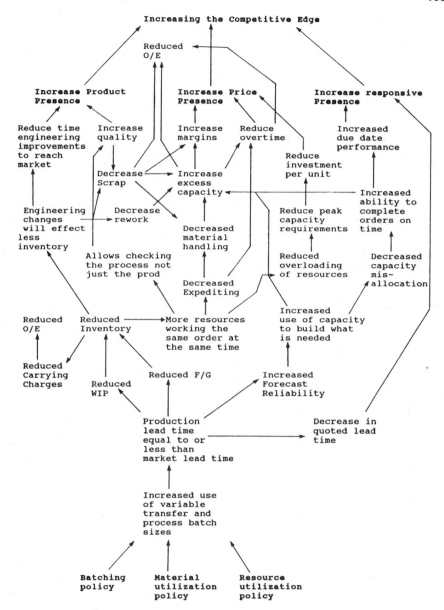

Figure 9.2 The relations diagram.

Each concept may be better distinguished from others through the use of boxes or circles. A double box or circle is used to set aside key problems or issues. (In Figure 9.2, bold type is used.)

The Tree Diagram

The tree diagram is a methodical process for describing the stepped activity necessary for achieving the desired goal. It too is a group activity and maps individual prerequisites that must be performed in order to accomplish a specific goal. To create the tree diagram, the following steps should be followed:

1. To adequately identify the goal to be accomplished, it is very important to understand precisely what the goal is. As in the relations diagram, cards or a blackboard may be used during the process to record ideas and to communicate.
2. In open forum, determine what the prerequisites are to accomplishing the goal and write them next to or under the goal with arrows or lines leading to the goal. Ask what must be done or what the prerequisite is to accomplishing the goal.
3. Once the first set of prerequisites has been finished, the next can begin. Starting with one of the prerequisites to the goal, determine what *its* prerequisites might be.

This process is repeated until all prerequisites have been discovered, whereupon it becomes the basis of an action plan for accomplishing the goal (Figure 9.3).

THE MATRIX DIAGRAM

The matrix diagram was designed to establish the extent of relationships or the degree of correlation between two sets of data such as:

- Tasks to be performed and the people or functions that must perform them
- Effects that exist in the environment and contributing causes
- Specific customer and design requirements

The matrix diagram can take many forms. The most commonly used is the L-shaped diagram in which one set of data is presented on the x-axis and another is presented on the y-axis. The correlation is determined at the point of intersection.

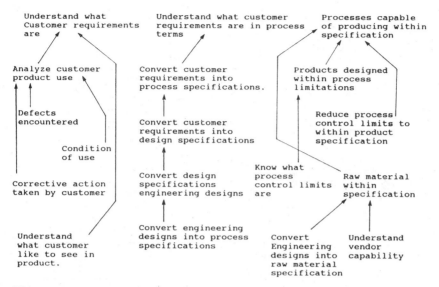

Figure 9.3   The tree diagram. The goal is to diagram processes capable of meeting customer requirements.

Figure 9.4 illustrates the use of the matrix diagram in assigning responsibility for prerequisites developed in Figure 9.3 (tree diagram).

One of the primary functions of the matrix diagram is at the very heart of quality function deployment. The process begins with an analysis of customer demand and proceeds through a series of matrix diagrams until the product and all its requirements have been fully described. Brainstorming sessions are used to aid in the process. Each matrix is used as input for the next in a cascading fashion which includes requirements of customers, product designs, product characteristics, manufacturing/purchasing, and control/verification (Figure 9.5).

In Figure 9.6 customer requirements are converted to design requirements and passed on to the next level, which details the engineering design requirements.

For a more definitive explanation of requirements each level may be broken down into primary, secondary and tertiary requirements (Figure 9.7).

A coding system is used to detail the relative importance of specific relationships between those at one level and the requirements at the next level down in the flow (Figure 9.8).

The relative importance of correlations is also established among requirements at the same level (Figure 9.9).

Function

P - Primary Responsibility C - Contributor Tasks	Design Eng	Manuf Eng	Quality	Production	Purchasing	Marketing	Cust. Serv
Analyze customer product use	C		C			P	P
Understand what customer wants	C					P	P
Convert cust requirement to design specifications	P		C	C		C	C
Convert design specs to engineering designs	P						
Convert engineering designs to process specs	P	P	C	C			
Provide process control limit information			P	P			
Design product w/in process control limits	P	C	C	C			
Reduce process variation to within product spec			P	P			
Convert eng. designs into raw material specs	P	C					
Understand vendor capability	C	C	C	C	P		

Figure 9.4 The matrix diagram.

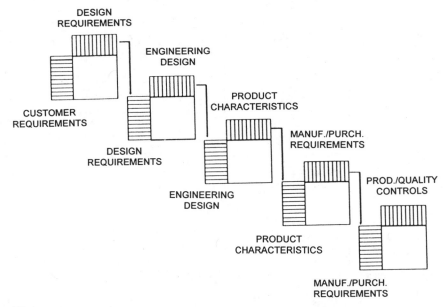

Figure 9.5 The QFD matrix.

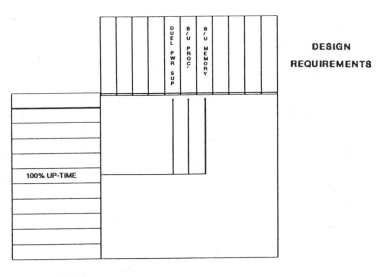

Figure 9.6 The design matrix.

PRIMARY	SECONDARY	TERTIARY

Figure 9.7 Using primary, secondary and tertiary functions.

⊙ VERY STRONG RELATIONSHIP

○ STRONG RELATIONSHIP

△ WEAK RELATIONSHIP

PRIMARY	SECONDARY	TERTIARY								
			○	○	○	○	○	○	⊙	○
			○	⊙		△	△	⊙		○
			⊙	△	⊙	△	⊙	○	○	○
			⊙	△	△	⊙		△		
			⊙	△			○			○
			△	⊙		⊙		○	△	△
					⊙				○	

Figure 9.8 Attaching relative importance.

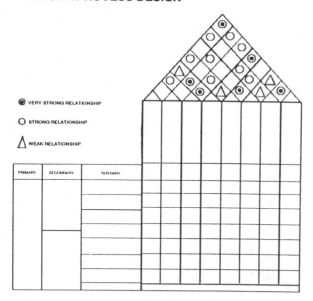

● VERY STRONG RELATIONSHIP

○ STRONG RELATIONSHIP

△ WEAK RELATIONSHIP

PRIMARY	SECONDARY	TERTIARY

Figure 9.9 Comparing importance at the same level.

The objective of the weighting system is to give relative priority to supporting activities and to establish the level of effort required. Additional information is added such as customer evaluations of competitive products, benchmarks, the degree of technical difficulty, the relative values of each design requirement in meeting the objective and the target values established during the engineering process. This process results in a total view of the relative quality of products and the priorities and technical difficulties associated with the project (Figure 9.10).

Improving the QFD Process by Applying TOC

Originally included in the QFD process (Hauser and Clausing, 1988) were provisions for including an estimation of relative cost (as a percentage) for each level in the process. But as this distorts the design process, another method must be found.

It is suggested that a better global perspective can be obtained by understanding the impact of certain design requirements and product characteristics on the three measurements of Throughput, Inventory and Operating Expense. Insight into this process is given in the section on constraint exploitation (p. 183). At issue is whether or not this kind of data can be included in the QFD

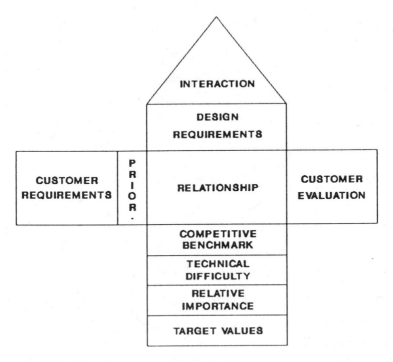

Figure 9.10 The house of equality.

process without distorting the relative importance of each characteristic from a global perspective.

Two measurements that may be used within the house of quality matrix are the relative cost of raw material and the degree of constraint absorption. Since minimizing each during the design process will result in more money coming into the company, analyzing each can prove to be an effective mechanism. This would require that two additional lines be added underneath target values in Figure 9.10. Each characteristic would be gauged (as a percentage) based on its relative impact on the constraint and total raw material cost. (Note that there are some problems in using this method that concern the aggregation of demand; these will be discussed later in more detail.)

Other matrixes not included in this brief discussion are T-type for relating two sets of data to a third; C-type, for correlating three data sets to an intersection point; the Y-type, for comparing three data sets to each other; and X-type, for correlating four data sets.

CUSTOMER REQUIREMENTS	P R I O R .	RELATIONSHIP	CUSTOMER EVALUATION
		COMPETITIVE BENCHMARK	
		TECHNICAL DIFFICULTY	
		RELATIVE IMPORTANCE	
		TARGET VALUES	
		RAW MAT. COST %	
		CONST. ABSORPTION %	

The Matrix Data Analysis Chart

The matrix data analysis chart is used for establishing the significance of relationships among variables and is used extensively in marketing research. A simple version may compare different types of products with different desirable attributes so that a visible comparison can be made between competitive products. In the following example the attributes of Great Taste and Less Filling appear on the x- and y-axes. Various customers would be asked to compare each beer against the two attributes. Each beer is coded to prevent bias.

```
                  Great Taste

  * C                    │ +3              *A
              *D         │ +2           *B
                         │ +1    +2    +3
  ─────────────────────────────────────────── Less Filling
      -3     -2     -1   │
                         │      *F
                    -2   │
        *E               │
                    -3   │
```

This type of information could be used to create a marketing or quality improvement plan for one of the beer companies.

The Process Decision Program Chart (PDPC)

The process decision program chart supports the creation of solutions through an analysis of possible alternatives. Like the tree diagrams, it offers a methodical and detailed approach to the thinking process. It begins with a problem or possible solution and attempts to predict the outcome. Unlike the tree diagram, where the question is, "What is the prerequisite?" the PDPC asks "What is the possible outcome or problem?" To create the PDPC, the following steps are followed:

1. Begin with a problem statement or possible problem solution. The tree and relations diagrams are excellent sources.
2. Ask the question, "What is the possible outcome or problem?"
3. After establishing an initial outcome, continue asking the same question until each branch has been completed.
4. Whenever a problem is encountered within a branch, list (to one side or below) countermeasures for overcoming the problem. Each counter-measure may be the subject of a separate PDPC.
5. Select the next problem or possible solution and continue the process.

Process decision program charts are excellent for use in areas where there is no experiential information available and for developing failure mode, effects and criticality analysis information (Figure 9.11).

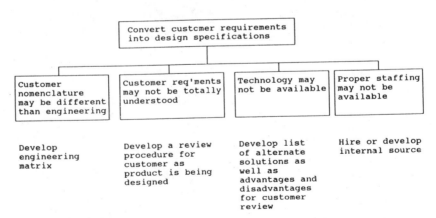

Figure 9.11 The process decision program chart.

The Arrow Diagram

The arrow diagram is used for determining and scheduling subordinate activities much like the program evaluation and review technique (PERT) or the critical path method (CPM). It can be used in conjunction with projects such as product design or construction. The following notation is used.

- *Arrows* indicate the direction of flow of a task from one node to another and the relative length of time required to perform a task.
- *Nodes* are indicated with circles and represent the starting or finishing of a task.
- *Numbers* indicate the order in which each node is placed.
- *Dotted lines* indicate connections between nodes but are not associated with time.

To create the arrow diagram, the following steps are followed:

1. Determine all necessary tasks to be performed and write each task on a separate card.
2. Determine that tasks/cards proceed or are performed in conjunction with other tasks/cards and place them in sequence.
3. Locate those cards that, when placed in a series, form the longest chain and lay them out end to end. This will be used to judge the relative position of all other cards.
4. Locate all parallel paths and place them in relative position.
5. Numbering each card in sequence and estimate the time required to perform each task.
6. Establish a unit of length for measuring time and place nodes, arrows and dotted lines on the medium used to present the diagram. Add the numbers for each node.

By using arrows to show the relative time, the critical path will be generated automatically. The time earliest (T/E)—the earliest time that a process can begin or end—is determined by adding the time associated with each arrow, in succession, to the time for the previous node, beginning with the first node. The T/E for the first node is time 0. The T/E for the second node, shown in Figure 12, is time 0 plus 5.

To determine the time latest (T/L), the latest time a process can begin or end, can be determined by starting at the last node (in Figure 9.12, node 6) and working backwards by subtracting the time required for the intervening arrow. When dealing with the critical path, the T/E and T/L will be the same. However, for those nodes not lying on the critical path, the T/L will be different (Figure 9.13).

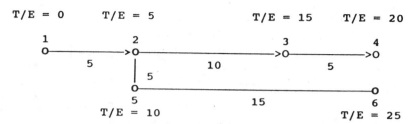

Figure 9.12 The arrow diagram.

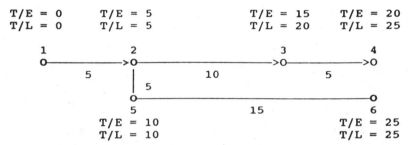

Figure 9.13 Establishing time latest (T/E) and time earliest (T/E).

The amount of "slack time," or allowable delay for any operation that can occur before jeopardizing the project's due date, is determined by subtracting the T/E from the T/L.

T/E = 0   T/E = 5   T/E = 15   T/E = 20
T/L = 0   T/L = 5   T/L = 20   T/L = 25
Slack = 0   Slack = 0   Slack = 5   Slack = 5

    1   2   3   4
    O————————>O——————————>O—————————>O
        5 | 10   5
               5
            O———————————————————————O
            5   15   6
        T/E = 10   T/E = 25
        T/L = 10   T/L = 25
        Slack = 0   Slack = 0

In Figure 9.13, the slack time for operations 3 and 4 is 5 days. The slack time for the critical path is 0.

Problems with the Arrow Diagram

Using the arrow diagram for scheduling projects may create more problems than it solves and may actually result in the extension of projects well beyond requirements for the following reasons:

- Like material requirements planning, it is capacity insensitive and reacts only to the order in which events should occur.
- It does not take into consideration the way in which resources interact, nor does it attempt to shorten lead times through maximizing the use of parallel operations.
- Since it is capacity insensitive, it cannot judge whether or not a constraint exists nor the extent of protective or excess capacity available.
- Lead times are stagnant, but in reality they are the result of the impact of the load at a specific resource or group of resources and, because they are dynamic, they cannot be predetermined. To determine lead times a schedule should be built first.
- Additional time allotted for "Murphy's Law" is spread throughout the project and cannot be judged based on its global impact.

A better method of scheduling projects is the DBR process (discussed in Chapter 6), which is based on the dynamics of interacting resources and dependent variables and is thus more effective in creating a valid schedule for maximizing Throughput. Care should be taken to ensure that operations are run in parallel whenever possible.

Additional Issues

When addressing planning issues and attempting to solve problems, it is imperative to ensure that core problems, and not just the symptoms, are addressed. Within the QFD umbrella the affinity diagram, relations diagram and matrix diagrams are excellent for creating correlations between data and for methodically analyzing complex issues. However, also needed is a method that increases the probability that core problems will be uncovered and solved. Many times relationships assumed to exist between activities in fact do not exist. Under TOC, whether satisfying the necessary condition or supporting the process of continuous profit improvement, the TOC thinking process (TP) increases the probability that core problems will be addressed by substantiating the supposed cause with effects or entities (see Chapter 4, on analyzing policy constraints).

Additionally, decisions during the design process, for example, between two alternatives or whether to maximize any given part specification, must be examined from more than just the customer's perspective. Chapter 11, on the impact of TOC on statistical process control, discusses the alternatives of scrap or rework on the Throughput of the company. These types of issues apply to the implementation of QFD at the resource level.

FAILURE MODE, EFFECTS AND CRITICALITY ANALYSIS (FMECA)

Failure mode, effects and criticality analysis is the methodical study of proposed product designs for possible failure conditions and their impact on the overall system and subsystems of products. The objective is to determine: (1) what product features are critical to a given mode of failure, (2) the effects of failure, and (3) what the possible solutions might be. Design efforts can then be focused on the most critical portion of the design based on the level of criticality. Critical components affect product safety, mission goals, manu-facturability and maintainability.

Block diagrams or tables are used to track the various cause-and-effect relationships and their criticality throughout production and to suggest alternatives (Figure 9.14).

The impact of the failure is given a rating from 1 to 10 based on the probability of occurrence, the seriousness of the results and how easy the failure would be to detect. A rating of 1 in each category would indicate that the problem

Product/ Feature	Failure Mode	Failure Effect	Failure Cause	Impact P S D	Possible Solutions
Tail Rotor Drive Shaft	Out of Balance	Med Speed Vibration	Loss of Balance Weight	5 5 2	Pre-Flight/ Maint/In- Flight Test Change Epoxy
	Loose Mounts	Med Speed Vibration	Metal Fatigue	3 9 2	Pre-Flight/ Maintenance Inspections Harden Mount
	Worn Bearings	Med Speed Vibration	Metal Fatigue	4 9 5	Pre-Flight/ Maintenance Inspections Harden Bearing

Figure 9.14 Failure mode effects and criticality analysis.

is not serious, that the probability of occurrence is low, and that the detectability is high. A rating of 10 would indicate that design of particular feature needs serious reconsideration.

	1 . . . 5 10	
Probability	Low	High
Seriousness	Low	High
Detectability	High	Low

The criticality index is determined for each failure mode by multiplying the rating for each category of impact.

Probability × Seriousness × Detectability = Index

The resulting index scores are then listed from highest to lowest to determine the order of priority. For the failure modes listed above the index would be:

Failure Mode	Failure Effect	Failure Cause	Rating
Worn Bearings	Med Speed Vibration	Metal Fatigue	180
Loose Mounts	Med Speed Vibration	Metal Fatigue	54
Out of Balance	Med Speed Vibration	Loss of Balance Weight	50

ANALYSIS OF RELIABILITY

The objective of reliability analysis is to predict the reliability of a system. There are two issues: the impact of time and the cumulative impact of component dependence

In analyzing the impact of time, the objective is to determine the probability of a product's functioning at specific points in time. The formula for product reliability at a specific point in time is

$$\text{Reliability (t)} = \frac{\text{The number surviving at time (t)}}{\text{Total number}}$$

If 100 parts are created and 70 are still functioning 12 months later, the reliability at 12 months would be

$$R(12) = \frac{70}{100} = 70\%$$

Reliability over time usually follows a predictable pattern in which products fail at an increasing rate at the beginning and end of their normal life cycles. The early stage is usually referred to as the "infant mortality" stage whereas the later stage is referred to as the "wearout" stage. The period in between is the "adult" stage.

In predicting system reliability, the interdependence of components must be considered. Each component is rated for individual reliability, and then the impact of using these components together is determined by multiplying successive reliability ratings together.

```
        Component      Rating

            A           90%
            B           90%


        A                   B

    ─────┤  R1  ├────────┤  R2  ├─────
```

```
Reliability  =  R1  X  R2
Reliability  =  .90  X  .90  =  .81 or 81%
```

The individual ratings for electronic components A and B are set at 90%. When these components are used together in a circuit, the rating would be 81% reliability. Reliability can be increased by using redundancy such that more than one part or group of parts performs the same function.

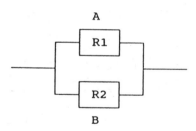

In this case parts A and B form a redundant circuit. The reliability for the circuit would be

```
Reliability  =  1-(1-R1)(1-R2)

Reliability  =  1-(1-.90)(1-.90)  =  .99 or 99%
```

Individual component reliability may be expressed in ways other than percentages, such as failures per thousand hours of use. In this case, the circuit's reliability would also be described in failures per thousand hours of use.

DESIGNING FOR MAINTAINABILITY

The importance of process maintainability in manufacturing is affected by the way that resources interface and by their availability to perform the mission. Availability is described as a function of mean time between failure (MTBF) and mean time to repair (MTTR). However, maintainability from a design perspective is generally customer driven and is more of a question of the allowable MTTR and the probability that repair will be executed within the allowable downtime.

Maintainability prediction assumes that MTTR is predictable, much the same as reliability and as such institutes a program for establishing component downtime probability and then developing an overall system for predicting mean time to repair. Historical information about repair time is gathered for like components of previous products and is assigned to current product designs. When the compiled information is compared with QFD requirements of availability and reliability, a clear picture of where improvements must take place develops. Based on historical data, the probability for repairing the product within certain time parameters can also be developed.

PHASED DEVELOPMENT

Phased development consists of the following stages:

- *Feasibility.* During the feasibility phase a determination is made as to whether the product can be designed and manufactured to the customer's needs and what the future impact will be on Throughput, Inventory and Operating Expense.
- *Design.* During the design phase, various alternative designs are reviewed and a primary candidate selected. Key issues include raw materials availability and cost, the impact on internal resources, and expected reliability and maintainability. Specifications are prepared in enough detail to create the prototype.
- *Prototype.* In the prototype phase, the product is built and tested to specification. Prototyping should accomplish two tasks: to test the design to see if the product performs as expected and to test the feasibility of manufacture.
- *Preproduction.* An additional, more advanced prototyping may need to be accomplished to iron out the details of full-scale production. The preproduction phase is used to "hand off" the product from design engineering to production. Manufacturing engineers review product designs as well as the prototypes to determine the best method of manufacture and request design changes, if necessary, to enhance manufacturability. Key issues include standardization of materials, degree of constraint absorption and fixture requirements.
- *Production.* This phase represents full-scale production and delivery to the customer. Special care is made to ensure that initial full-production problems are addressed.

An additional phase is sometimes added for postproduction design review to ensure continued manufacturability and to suggest product improvements.

DESIGN REVIEW

Each phase of development is accompanied by a design review. The design review team ensures that a wide variety of expertise is available to review each phase of design so that the problems that normally accompany the introduction of a new product are minimized. The design review team should ensure that designs are reviewed well ahead of the production release date and that each meeting is well planned so that documentation is available. Checklists are usually used for the organization of the meeting so that all aspects of the product are reviewed.

Special care must be given to creating a nonthreatening atmosphere for the review. The meeting should be positive, rather than a chance to "hang the designer." Figure 9.15 is a modification of the design review team membership and responsibility chart taken from Juran's *Quality Control Handbook* (Juran, 1983) and outlines the responsibilities of individual design team members. Special consideration must be given to issues of invalid cost management as well as to the impact of the design process on the dependent-variable environment and the over-all profitability of the company. Asterisks in Figure 9.15 indicate modifications.

Every member of the design review team must be keenly aware of what actions are necessary in the design process to ensure the overall profitability of the company and the impact of their actions on the dependent-variable environment.

EXPLOITING THE CONSTRAINT*

Perhaps the largest contributor to product quality is the concept of the manufacturability of design. Factors having a direct effect on manufacturability include the following American Production and Inventory Control Society, 1991):

Number of parts (minimize)
Variability of components (minimize)
Process requirements (simplify)
Degree of design detail (simplify)
Technology employed (should be proven)
Availability of materials (high)
Degree of production involvement (high)
Design for yield (within process capabilities)

While these are excellent guidelines for the design process in preventing problems from occurring, they themselves create constraints if not given the proper amount of attention. Two major areas of focus that have an immediate impact on the amount of Throughput being generated by the constraint are

- Product design/Redesign to minimize constraint resource time for each product being produced.
- Routing changes for products that can be produced in a different way or at different resources.

In the product mix decision process discussed in Chapter 5, it was discovered that the amount of Throughput generated for the constraint time used

---

*Reprinted from *Production and Inventory Management Journal 32*(1): 1991, p. 7, by permission of the American Production and Inventory Control Society.

Member	Responsibility
Chairperson	Calls, conducts meetings of group, and issues interim and final reports.
Design Engineer	Prepares and presents designs and substantiates decisions with data from tests or calculations.
Reliability Engineer	Evaluates design for optimum reliability consistent with goals.
Quality Engineer	Insures that the functions of inspection, control, and test can be effectively carried out.
Manufacturing Engineer	Insures that designs are produceable, reviews the impact new designs will have on internal resource constraints/non-constraints, that constraint utilization is minimized where possible and that subordination can be properly accomplished.
Field engineer	Insures that installation, maintenance, and user considerations were included in the design.
Purchasing	Insures that acceptable parts and materials are available to meet cost and delivery schedules.
Materials Engineer	Insures that materials selected will perform as required.
Tooling Engineer	Evaluates design in terms of the tooling impact on T, I & O/E as well as tolerance and functional requirements. Special consideration is given to the overall impact of the new design on the dependent variable issues. As an Example, should a new tool be created which will reduce the impact the new product will have on the constraint and therefore increase the Throughput created per unit of the constraint.
Packaging and Shipping Engineer	Assures that the product can be safely handled without damage.
Marketing Representative	Assures that requirements of customers are realistic and fully understood by all parties. The marketing rep. is also there to understand the impact the new product will have on current marketing strategies. As an Example, will the new product change the location of the constraint and therefore the pricing, product mix and marketing segmentation strategies and, if so, how.
Design Engineer (not associated with product under review)	Constructively reviews adequacy of design to meet all requirements of customer.
Consultants, specialists on components, value, human factors etc.	Evaluates design for compliance with goals of performance, the impact on the dependent variable environment and schedule.
Customer Representative	Generally voices opinion as to the acceptability of design and may request further investigation on specific items.

Figure 9.15   The design review team. Asterisks indicate modification. (From Juran, 1988.)

in making each product determines which product mix will be the most profitable. The lower the amount of constraint time per Throughput dollar created, the higher the amount of Throughput generated for the company. Those products that have the highest Throughput rating per unit of the constraint are given the highest priority in absorbing constraint time. In the engineering process, whenever products are designed or redesigned, careful consideration must be given to ensuring that the amount of time for all products processed at the constraint is held to a minimum. As the ratio of Throughput per unit of the constraint goes up, the amount of money coming into the company also increases. Process changes must also consider this issue. If a routing change can be made that decreases the constraint time, then the amount of money that can be made will increase.

In the activity-based cost information presented in Chapter 5, the following information was used to determine the most profitable product mix:

Routing Part A				Routing Part B		
Res	Op.	Time		Res	Op.	Time
123	10	10		123	10	10
124	20	30		124	20	15
125	30	20		125	30	20

Res	Time Avail.	Demand A	B	Total	Delta
123	7200	1000	2000	3000	+4200
124	5400	3000	3000	6000	- 600
125	7200	2000	4000	6000	+1200

	Mat.	Lab.	Ovr. Hd.	Std Cost	Sales Price	Prof. Mar.	Quan. Sold	Cash Gen.
A	80	60	60	200	300	100	80	17,600
B	80	60	180	320	310	-10	200	46,000
								63,600
								-60,000
								3 600

	A	B
Sales Price	300	310
Raw Material	80	80
Cash Generated	220/30 = $7.3	230/15 = $15.3

Resource 124 was identified as the limiting factor because it needed 600 more hours to meet demand created by the market. Product A required 30 minutes of constraint time and generated $220, while product B used 15 minutes of constraint time and generated $230. The more profitable product was determined to be product B. For every minute of constraint time used for product B, $15.3 was created, as compared with $7.3 for product A. So the determination was made to use as much constraint time making product B as possible. Any time left over could then be used in making product A.

If the routing for product B were to change or if a modification were made that reduced by one minute the amount of time it took to make product B, the result would be an increase in productive capability, resulting in 200 minutes (one minute for every product B made and sold) being made available to make product A. The increase in Throughput would be immediate. An additional 6 units of A could be made, which would result in an increase of $1320.

	Mat.	Lab.	Ovr. Hd.	Std Cost	Sales Price	Prof. Mar.	Quan. Sold	Cash Gen.
A	80	60	60	200	300	100	86	18,920*
B	80	60	180	320	310	-10	200	46,000
								64,920
								-60,000
								4,920

Any time saved at the constraint will almost surely result in an increase in throughput. The process of determining where to focus the engineering effort begins with a Pareto analysis of demand. What products/operations use the most constraint time based on the current load? Part A uses 30 minutes of constraint time and 60 minutes of total resource time, while part B uses 15 minutes of constraint time and 45 minutes of total resource time. It would seem that the best place to look for improvements would be part A. However, the largest constraint load is actually created by part B, with a total demand of 3000 minutes compared to 2400 minutes used by part A. In addition, each minute saved at part/operation B/20 will result in a total of 200 minutes being released and which can be used in making more units of A to meet market demand. Figure 9.16 shows a Pareto analysis of parts/operations that cross the constraint.

While not all improvements originate from first going through a Pareto analysis of part/operation demand, the Pareto analysis remains an excellent starting point in determining where to focus engineering improvements. Additional considerations include the reliability or capability of a given product or process. Obviously, if a product is difficult to make at the constraint, it will take up more

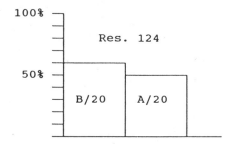

		Product	B
		Part Operation	B/20
		Time per part	15min.
		Total time	3000min

		Product	A
		Part Operation	A/20
		Time per part	30min.
		Total time	2400min

Figure 9.16 Exploiting the constraint.

constraint time. The more time spent at the constraint, the less opportunity to create Throughput.

SUMMARY

Meeting the necessary conditions of the customer while designing products to increase productivity and protect the creation of Throughput is the objective of the product/process design system under TOC-based TQM. It is imperative that designers become aware of their role in supporting the overall effort to increase profits.

STUDY QUESTIONS

1. What is the purpose of the design function?
2. Define the term "quality function deployment" (QFD).
3. List and define the "seven new tools" used in the implementation of QFD.
4. What concepts are missing from the traditional implementation of the QFD process?
5. What "new tool" is used in the "house of quality," and what key issues are overlooked in this process?
6. What problems are associated with the arrow diagram?
7. What is meant by the term "mean time between failures" (MTBF)?
8. What is meant by the term "mean time to repair" (MTTR)?
9. List and define the five steps of phased development
10. What is the purpose of the design review?

11. What design efforts should be considered to support the five steps of continuous profit improvement?
12. What possible effects can be expected from reducing the amount of constraint time absorbed by a given product?

10

Statistical Process Control:
An Introduction

This chapter is designed to provide an introduction to the process of statistical process control. It is a prerequisite to Chapter 11, "TOC and Statistical Process Control." For a complete discussion of SPC under the traditional approach, see Doty (1991).

OBJECTIVES

- To introduce the concepts of variability and process capability
- To demonstrate the basic use of charts and the charting process.
- To show how basic charts are constructed.
- To explain how to interpret charted data and devise solutions.

NORMAL VARIATION

Normal variations within a process are predictable. When plotted as a graph, the frequency of occurrences describes what is normally referred to as a bell-shaped curve. The predictability of the normal distribution, or curve, can be used to determine whether a process is in control or is capable of creating a defect-free part. When applied to manufacturing processes, statistics can be used to predict the size, or distribution of sizes, of different parts produced from specific resources.

In a bell-shaped curve, the majority of occurrences appear around the average (center of the distribution, or mean) (Figure 10.1). Notice that the farther away from the mean, the fewer the number of occurrences.

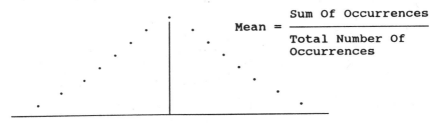

Figure 10.1 The bell-shaped curve.

If the total range of the normal curve is divided into 6 equal segments denoted by sigma (σ), the majority (68.3%) of all measurements should fall within $\pm 1\sigma$.

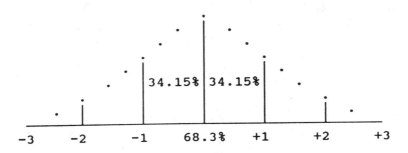

In other words, 34.2% of the measurements taken fall between the mean and -1σ, and 34.2% should fall between the mean and $+1\sigma$. The remainder of the distribution should resemble Figure 10.2, in which 99.7% of all measurements should fall within $\pm 6\sigma$. If this is not the case, then it is likely that the process is not in control.

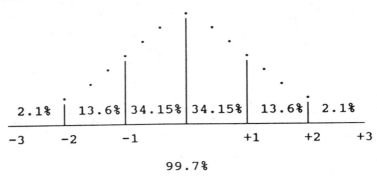

Figure 10.2 Sigma in the normal curve.

Finding σ values for a process is not as easy as dividing the total range by 6. Values of σ indicate the probability that a number will fall within a certain measurement range. The smaller the sample size, the lower the confidence one can have of being correct.

Statistical tables are used to designate multipliers for finding σ based on the sample size. The formula for σ_x reads as follows.

$$\sigma = \frac{\overline{R}}{D_2}$$

$\overline{R}$ is the average range of measurements taken. A sample size of 5 parts taken at different times yields an average range. The range is equal to the highest value minus the lowest value within the same subsample.

8:00	9:00	10:00	11:00	12:00
8	5	10	5	8
7	7	6	9	4
6	9	9	4	7
5	4	4	8	6
4	6	7	7	5

The range for the 8:00 sample is equal to 8 minus 4. The ranges for all samples are averaged to arrive at the average range, which in this case is 4.8, which is rounded to 5 as noted (*) below and on p. 199.

$$\sigma = \frac{\overline{R}}{D_2} = \frac{4.8}{2.326} = 2.06$$

Sample size	D_2
2	1.128
3	1.693
4	2.059
* 5	2.326
6	2.534
7	2.704

PROCESS CAPABILITY

Once an understanding of what a resource should be able to produce on a consistent basis is determined through statistical means, it must also be determined
whether the resource can produce within engineering specifications.   In
determining whether a process is capable of producing parts consistently within
engineering specifications, the specification range is divided by the capability of
the process.   The specification range is determined by the engineering
specification. Process capability is determined by multiplying σ for the process by
6. As an example, the specification for a ¼-inch bolt is .25 inches ±.015. The
range would be from +.015 to −.015, or .030 inches. If σ for the process were
.00375, 6σ would be .00375 × 6, or .0225. The capability index (C_p) for the
process would be:

$$Cp = \frac{\text{Specification Range}}{\text{Process Capability}} = \frac{.030}{.0225} = 1.33$$

Note that with a C_p of less than 1.33, the process is considered incapable of
consistently producing defect free parts.

In this case, the mean of the process and the midpoint of the specification or
target value are both .25 inches. The process capability and the specification are
graphically represented as follows.

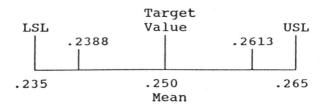

```
                      Target
   LSL                Value              USL
        .2388                 .2613
    |     |            |        |         |
    |     |                     |         |

  .235              .250                 .265
                    Mean
```

What would be the impact if the mean of the process were to shift in either the
plus or minus direction while the 6σ spread remained the same? There would be
an increased probability of defects appearing at the end of the specification toward
that the shift was made.

```
                      Target
   LSL                Value              USL
         .2407                 .2632|
    |      |           |      |       |
    |      |                  |       |

  .235               .250             .265
                     Mean
                     .252
```

To ensure that the capability index reflects this change, the C_pk index is used. To find the C_pk index, (1) find Z_{min}, then divide Z_{min} by 3. Z_{min} is found by first finding Z_{upper} and Z_{lower}. The lower of the two is Z_{min}.

$$\text{Z upper} = \frac{|\text{Upper Specification Limit - Mean}|}{\text{Sigma}}$$

$$\text{Z lower} = \frac{|\text{Lower Specification Limit - Mean}|}{\text{Sigma}}$$

Notice that the numerator in the equation is an absolute value. In other words any negative numbers are converted to positive numbers.

$$\text{Cpk} = \frac{\text{Z min}}{3}$$

Z_{upper} for the process described above is

$$\text{Z upper} = \frac{.265 - .252}{.00375} = 3.46$$

Z_{lower} for the process described above is

$$\text{Z lower} = \frac{.235 - .252}{.00375} = 4.53$$

Z_{min} is the lower of the two and is

$$\text{Z min} = 3.46$$

C_pk is

$$\text{Cpk} = \frac{3.46}{3} = 1.15$$

Since the mean of the process shifts to the right, the C_pk drops from 1.33 to 1.15, and the process is no longer considered capable of consistently producing defect-free parts.

QUANTIFYING THE EXPECTED LOSS

To determine the number of defective parts expected at either limit of the specification, Z_{upper} and Z_{lower} are compared against a table for determining the area under the normal curve (see Table A.1 in the Appendix). In the case above, Z_{upper} is 3.46. By finding 3.4 on the table under Z, moving across to .00027 under 0.06, and then moving the decimal two places to the right, the percentile figure is determined. For a Z_{upper} value of 3.46, a rejection rate of .027% can be expected. A rejection rate for Z_{lower} can be found in the same manner. A Z_{lower} value of 4.53 will result in a .00029% rejection rate. When added together, the total rejection rate expected is .02729%.

In Appendix A.1, the single numbers just to the right of the Z column indicate the number of zeros (followed by a decimal point) that should appear to the left of the entry. As an example, when finding the expected loss for Z_{upper}, 3.46, three zeros are added to the left of 27009 and the last three digits are rounded up. For Z_{lower}, five zeros are added to the left of the entry.

It is important to determine which specification limits (upper and lower) are associated with rework and which are associated with scrap. The global impact of this is discussed in Chapter 11, on implementing SPC for quality and profit.

X-BAR AND R CHART CONSTRUCTION

The first step in the charting process is to construct the chart. Located at the top of most charts is administrative information, which may include the part name and number, the operation, the name of the operator, specifications, the gauge to be used, and so forth (Figure 10.3).

Part Number	Nomenclature			
12345	Armature Flying Grammis			

Operator	Machine/Op.	UOM	Zero Base	Engineering Spec.
John	LATHE 123	.001	.500	.500 – .515

Figure 10.3 The X-bar and R charts.

Specification limits refer to the engineering specification. The zero base is a number used in the coding process. It is not always desirable to maintain exact copies of measurements. To simplify the process, coding is used. Coding involves several different methods for expressing data but basically the subtraction of a constant from actual data to create a smaller number. When the constant is added back, the original number is re-created. Zero base numbers for coding include:

- *Common number*: A number that is common to all numbers in the data base data.
- *Target value*: The midpoint of the engineering specification.
- *Smallest measurement*: The smallest measurement within the data base.
- *Largest measurement*: The largest measurement within the data base.
- *Nice round number*: Any number that can be easily subtracted from the data collected.
- *Upper or lower specification limit.*

As an example, the numbers .510, .509, .508 and .504 represent actual measurements. To reduce these numbers for easier manipulation, establish the zero base and subtract. If the zero base were set at .500, then the coded numbers would be 10, 9, 8, and 4. If it were set at the lowest measurement (.504), the coded numbers would be 6, 5, 4, and 0. When the lowest measurement is added to 6, the "restored" number is .510.

Charts also contain an area to record measurements and to compute range and average figures. The bottom half is used to plot measurements graphically and to set certain control mechanisms. Once the zero base has been established and the sampling plan has been set, data collection can begin. The initial objective is to collect enough information to determine the average mean and average range for a particular operation and to create the centerline for the average and range sections in the bottom half of the chart. At least 20 measurements are necessary to do this. The sampling plan and data used in the capability section are still applicable.

8:00	9:00	10:00	11:00	12:00
8	5	10	5	8
7	7	6	9	4
6	9	9	4	7
5	4	4	8	6
4	6	7	7	5

Time	8:00	9:00	10:00	11:00	12:00
S M	8	5	10	5	8
a e	7	7	6	9	4
m a					
p s	6	9	9	4	7
l u					
e r	5	4	4	8	6
e					
s	4	6	7	7	5

Figure 10.4 The measurement samples.

The plan is to inspect five consecutive parts from the process every hour. At the end of five hours, 25 parts have been inspected. The measurements are entered on the chart under the applicable time in the sample measurements section (Figure 10.4).

Directly under the sample measurements section is a section for entering the sum, the average and range for the measurements taken. The sum is determined by adding the measurements. The mean is then determined by dividing the sum by subgroup sample size. The range is determined by subtracting the highest measurement from the lowest within subgroup samples. Figure 10.5 shows the results of the measurements from Figure 10.4.

The notes section is used to enter any process changes and may give clues for determining the causes of any changes in the measurements that indicate an out-of-control situation.

Sum	30	31	36	33	30
Average, X	6.0	6.2	7.2	6.6	6.0
Range, R	4	5	6	5	4
Notes					

Figure 10.5 Establishing the sum, average and range.

Before one can fill out the average and range plotting sections, the grand average and average range must be determined. The grand average is determined by finding the sum of all the averages from left to right and then dividing by the number of averages used. The average range is determined by finding the sum of all the ranges and dividing by the number of ranges used. From Figure 10.5, the following grand average and average range are computed.

$$\text{Grand Average } (\bar{\bar{X}}) = 6.0 + 6.2 + 7.2 + 6.6 + 6.0 = 32/5 = 6.4$$

$$\text{Average Range } (\bar{R}) = 4 + 5 + 6 + 5 + 4 = 24/5 = 4.8$$

The grand average $(\bar{\bar{X}})$ and the average range $(\bar{R})$ are used to determine the average and range centerlines for the chart in the plotting section. Once the values are determined, the centerlines can be plotted.

Sum	30	31	36	33	30
Average, X	6.0	6.2	7.2	6.6	6.0
Range, R	4	5	6	5	4
Notes					

A
v
e
r $\bar{X}$ 6.4
a
g
e

R
a
n R   4.8
g
e

After the center lines are created, control and warning limits are set so that out-of-control conditions can be easily detected and causes interpreted. Control limits are set at ±3σ from the grand average or from the average range. Warning limits are set at ±2σ. To set control and warning limits, the following formulas are used:

Average ($\bar{\bar{X}}$)

$$\text{Upper Control Limit (UCL}_{\bar{X}}) = \bar{\bar{X}} + A_2 \bar{R} = 6.4 + (.577 \times 4.8) = 9.17$$

$$\text{Lower Control Limit (LCL}_{\bar{X}}) = \bar{\bar{X}} - A_2 \bar{R} = 6.4 - (.577 \times 4.8) = 3.63$$

$$\text{Upper Warning Limit (UWL}_{\bar{X}}) = \bar{\bar{X}} + 2 \left(\frac{A_2 \bar{R}}{3}\right) = 6.4 + 2 \left(\frac{.577 \times 4.8}{3}\right) = 8.25$$

$$\text{Lower Warning Limit (LWL}_{\bar{X}}) = \bar{\bar{X}} - 2 \left(\frac{A_2 \bar{R}}{3}\right) = 6.4 - 2 \left(\frac{.577 \times 4.8}{3}\right) = 4.55$$

Range ($\bar{R}$)

$$\text{Upper Control Limit (UCL}_R) = D_4 \bar{R} = 2.115 \times 4.8 = 10.15$$

$$\text{Upper Warning Limit (UWL}_R) = 2 \left(\frac{D_4 \bar{R} - \bar{R}}{3}\right) + \bar{R} = 2 \left(\frac{2.115 \times 4.8 - 4.8}{3}\right) + 4.8$$

$$= 8.37$$

Figure 10.6 represents the table of factors for computing control and warning limits.

Subgroup Size	A 2	D 4
2	1.880	3.267
3	1.023	2.575
4	.729	2.282
* 5	.577	2.115
6	.483	2.004
7	.419	1.924

Figure 10.6 Table of factors.

Once the upper and lower control and warning limits have been determined, these should be shown graphically with heavy dotted lines so that it can be easily determined if a problem exists.

PLOTTING

After the chart has been prepared, the plotting can begin. The numbers to be plotted are from the average and range for each subgroup. For each subgroup, a separate plot is made using as a reference the centerline established earlier and the time that the subgroup sample was taken (Figure 10.7).

SPC CHART INTERPRETATION

Chart are used to determine whether a process is under control, and if it is not, what might be causing out-of-control conditions.   In both the average and

Time	8:00	9:00	10:00	11:00	12:00
S M	8	5	10	5	8
a e m a	7	7	6	9	4
p s l u	6	9	9	4	7
e r e	5	4	4	8	6
s	4	6	7	7	5

	8:00	9:00	10:00	11:00	12:00
Sum	30	31	36	33	30
Average, X	6.0	6.2	7.2	6.6	6.0
Range, R	4	5	6	5	4
Notes					

Figure 10.7 Plotting the data.

range charts below, under normal variation, at least 68.3% of all measurements should fall within ±1σ of the centerline measurement, at least 95.5% within ±2σ and at least 99.7% within ±3σ. If this not the case, something is wrong with the process. A distribution that does not meet these parameters indicates an abnormal variation and that something is wrong with the process.

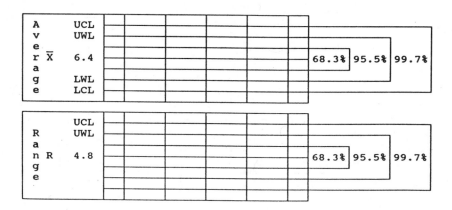

Abnormal patterns that indicate that certain conditions may exist in the process are of several types:

- *Freaks.* Any one point outside 3σ on either side of the centerline. Major causes are usually human error in arithmetic, plotting or measurement. Also caused by material problems.

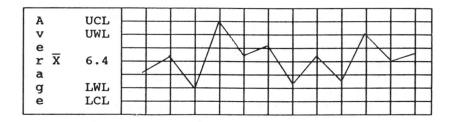

- *Freak patterns.* Two out of three measurements outside 2σ on either side of the centerline or four out of five measurements outside 1σ on either side of the centerline. Usually caused by material or measurement problems.

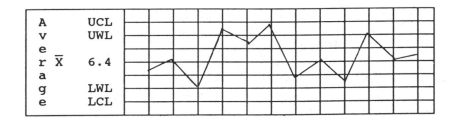

- *Shifts.* Seven measurements in a row on one side of the centerline. Usually caused by a change in material, machine speed or setup.

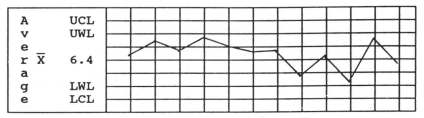

- *Trends.* Six measurements in a row, each above the last measurement, or six measurements in a row, each below the last measurement. Caused by a gradual change such as tool wear.

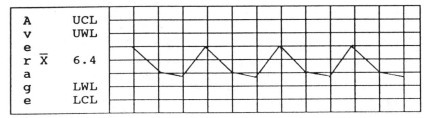

- *Cyclical patterns.* Patterns that repeat themselves. Usually caused by human factors such as shift changes. May also be caused by defective equipment.

- *Jumps.* A movement of 4σ or more between two consecutive measurements. Usually indicates that something has broken.

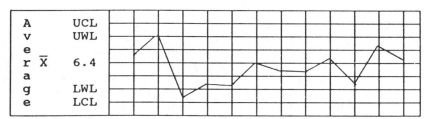

- *Stable mixtures.* Five points in a row outside 1σ on both sides of the centerline with no points within 1σ. Usually indicates the mixture of two different processes. One major cause is material related. Can also be caused by different operators or measuring equipment.

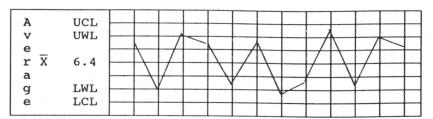

- *Clustering.* Measurements that occur in clusters all over the chart. Indicative of mixing materials from different vendors or processes. Causes can be human error, material, machine, method or tooling.

- *Erratic patterns.* With 10 measurements plotted, 40% are outside 2σ or 30% outside 3σ. Primary causes include operator error, overadjustment or inconsistent measurement

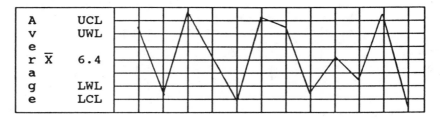

- *Stratification.* Fourteen points in a row inside 1σ. Caused by not measuring to a fine enough tolerance or by operators fudging the figures.

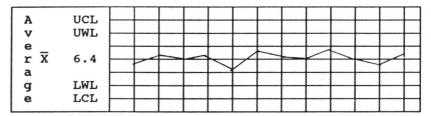

Once a pattern has been identified, a cause is then hypothesized and a solution implemented that will bring the pattern back into normal distribution. Tools for solving problems in out-of-control conditions include brainstorming to identify speculative causes and possible solutions. Fishbone analysis is used for organizing the process and for designing experiments when a large number of variables offer many alternatives.

ADDITIONAL CHARTS

The X-bar and R charts are usually used to track variable-type data, that is, data that is measured. In addition to variable-type charts, there are charts that record attributes data, which simply measure decisions as good or bad. In general, they are used to analyze the number or percentage of defects that occur. These additional charts include the p chart and its derivatives for controlling defective parts, and the c and u charts for controlling specific defects or the quality characteristics within the part.

• • •

The methods for implementing SPC and deciding where to focus have undergone tremendous change.

• • •

STUDY QUESTIONS

1. What is the objective of statistical process control (SPC)?
2. In what way(s) can SPC be used to support the goal of the company?
3. Define normal variation and why it occurs.
4. What is the impact of normal variation on SPC?
5. Explain the concept of σ and how it is used.
6. Explain the concept of capability indexing and how a resource is determined to be capable?
7. What are Z_{upper} and Z_{lower}?
8. List five methods for arriving at a zero base.
9. What is the grand average and what is it used for?
10. What is the average range and what is it used for?
11. What is the objective of SPC chart interpretation?
12. Define the following chart patterns and explains their causes: freak pattern, shift, trends, cyclical pattern, stable mixture, clustering, erratic patterns, stratification.

11

TOC and Statistical Process Control

This chapter is dedicated to understanding how to properly implement SPC to support the necessary condition of good quality and the process of continuous profit improvement.

OBJECTIVES

- To establish the relationship between SPC and the way in which re-sources interface
- To understand those areas within the production process that have the greatest impact on the implementation of SPC
- To understand how the SPC program is used in the exploitation and subordination phases
- To create the relationship between the SPC program and customer demand
- To establish the relationship between the SPC implementation and its impact on the decision process
- To establish the implementation procedure and to address those actions that could it

Under TOC, SPC is used to continuously support the exploitation and subordination phases within the five-step improvement process so that Throughput can be enhanced or protected and to support the QFD program in meeting the necessary condition. To do this, the implementor must know where to focus, how to implement SPC and how to prevent SPC from being blocked.

FOCUSING THE SPC EFFORT

The global impact of an improvement cannot be ascertained based on a localized measurement such as rejection rates. To understand what is important in implementing SPC, a method must be used to determine the relationships among resources. To do this, the product flow diagram is used (Figure 11.1).

As discussed in Chapter 3, the product flow diagram is a detailed description of the flow of product from each gating operation to the sales order. Included in the basic diagram are:

- The bill of material, designating the part number and determining product levels
- The routing, designating the sequence of operations
- Resource information, designating where an operation is to be performed

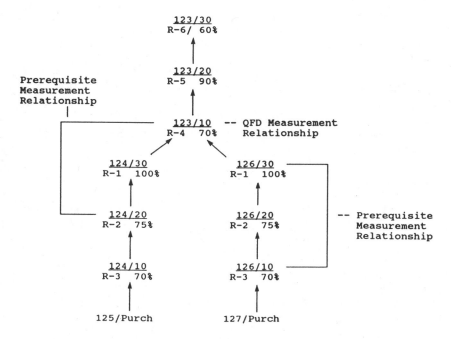

Figure 11.1 The product flow diagram.

As shown in Figure 11.1, example resource load information has been added, giving an indication as to the severity of a particular problem from a global perspective. Those resources that can least afford a problem are expected to accomplish the most. However, there are more subtle issues to be addressed:

- The string of resources leading from the constraint to the sales order
- Those resources whose characteristics are prerequisites for the constraint
- Those resources whose characteristics are prerequisites for QFD
- Those resources that create holes in the buffer because of scrap

In Figure 11.1, resource R-1 is considered the constraint since it is loaded to 100%. Resources R-4, R-5 and R-6 represent the string of resources leading from the constraint to the sales order. In processing orders from resource R-1 through R-6 great care must be used to ensure that no scrap occurs. Products being processed at or that have already passed the constraint are considered more valuable than those that have not. Constraint time governs the amount of Throughput that can be generated. Its limited availability means that no additional parts can be made without sacrificing Throughput. Also, resource R-5 also has limited in capacity. Its being loaded to 90% means the chances are very good that the orders going through this resource will be late. It can ill afford rework as well. Resources R-4 and R-6 have additional capacity to handle rework.

While not originally presented in the product flow diagram an examination of product specifications finds that the prerequisites for creating acceptable measurements at part/operation 126/30, performed at resource R-1, are the measurements obtained at part/operation 126/10, performed at R-3. An over- or undersized measurement could result in scrap or rework at 126/30 (at resource R-1, the constraint).

Also not considered is the impact of QFD. An examination of the QFD matrix diagram shows a very strong relationship between the measurements obtained at part/operation 123/10, performed at resource R-4, and customer requirements. There has also been found a measurement relationship between measurements of 123/10, performed at R-4, and 124/20, performed at R-2.

What does all this mean to the implementation of SPC? The first implication is the sequencing of the implementation. Those resources having problems and that impact Throughput the most should be implemented first. If a scrap or rework problem is impacting the constraint, or would cause the necessary conditions required by 123/10 on R-4 to be missed, a study of the capability and implementation of statistical control of resource R-1 and R-4 would be highly recommended. The second implication concerns the improvement process. Those resources that restrict Throughput should be improved first.

There are also implications for those resources that are not capable but that must perform so that damage to Throughput is minimized. After examining the product flow diagram, it is discovered that certain resources, because of capacity, location and customer requirements are required to fulfill certain quality-related responsibilities including

- The prevention of scrap
- The prevention of any type of loss
- The support of QFD
- The support of other resources

Resources R-1 and R-5 are critical because of the load placed on them. Any loss from these two resources, whether rework or scrap, will result in either a permanent loss of Throughput or a delayed order. To minimize any loss the C_pk index should be 1.33 or better at these parts/operations/resources and all attempts must be made to keep them in control. Resources R-4 and R-6, while they can take a certain amount of rework, cannot afford scrap, and all attempts should be made to prevent rework. Work improvement team efforts should focus on any loss at R-1 and R-5 and any scrap at R-4 and R-6.

ADAPTING TO CAPABILITY

For conditions where the C_pk is less than 1.33, by moving the mean for the operation toward the rework side of the specification, scrap can be minimized. Since additional capacity is available to complete rework and still finish the order on time, Throughput would not be threatened. However, one may need to consider design implications as well as the impact on the capability of other resources. Products are designed to fit together around the nominal of the specification. If the mean of the process distribution is to be moved, the means of other resources as well as design specifications will need examining.

In certain cases, the constraint will have a low C_pk index and scrap or rework is unavoidable. The best alternative might be to move the constraint by elevating the constraint resource using whatever methods are available. This is not always possible. The issue now becomes how to minimize the amount of damage incurred. To be able to totally understand the result, the cumulative impact of scrap or rework must be known across the entire schedule. The key issues are

- The cumulative value of the time taken from the constraint in Throughput terms by rework or scrap
- The cumulative value of the raw material loss projection due to scrap
- The increase, if any, in Inventory or Operating Expense due to an increase in load requirements at non- or near-constraint resources existing prior to the constraint and caused by re-creating the replacement parts

To answer this question a valid schedule must be produced for both alternatives, and the impact determined.

THE IMPACT ON PRODUCT MIX

There are other considerations. A decision about how to maximize profitability, given the current limitations, must still be made. The additional time needed to process the part at the constraint because of rework or scrap will add to the total time used to process the part. This affects not only the utilization of constraint time, but also the choice of product mix (see Chapter 5). The amount of constraint time absorbed by the product is thus a key issue. (Figure 11.2).

The formula for the decision is:

$$\frac{\text{Throughput}}{\text{Total constraint time per unit}}$$

The amount of Throughput generated by parts A and B each time they are sold is $100. For product A, the amount of constraint time absorbed is 40 minutes, and for product B, the amount is 30 minutes, making product B (at $3.33 per constraint minute) more profitable. However, when rework is considered, product A becomes more profitable. B's total constraint absorption increases from 30 minutes to 60 minutes. And the Throughput per unit of the constraint declines to $1.66.

What has not been considered in this problem is the rate of rework for the current schedule. On average, the amount of rework may be 10%, or 3 minutes additional CCR time for all units of product B. Instead of using the full 30 minutes of rework time, the reduced value of 3 minutes (10% of 30 minutes) is used. Product B then becomes the more profitable.

	A	B
Total CCR Time per unit	40 min	33 min
Throughput per unit of the Constraint	$2.50	$3.03

	Product	(Rework) Product
	A	B
Throughput Generated	$100	$100
Original CCR Time per unit	40 min.	30 min.
Throughput per unit of the Constraint	$2.50	$3.33
Additional CCR Time per unit Due to Rework		30 min
Total CCR Time per unit	40 min	60 min
Throughput per unit of the Constraint	$2.50	$1.66

Figure 11.2 The impact on product mix. CCR = capacity-constrained resource.

With scrap, the situation is similar, except that the loss from having to replace the raw material must be included in the equation as well. The amount of Throughput generated per unit must be reduced by the raw material cost. Throughput for B (scrap) is computed as Throughput minus raw material. Throughput has been reduced from $100 to $75.

In this case, it would be more profitable to scrap the product than rework it since the Throughput per unit of the constraint is higher.

In considering process/product design in this case, it would be more profitable to set the mean of the distribution of measurements for the constraint resource to the scrap side of the specification or move the nominal value of the specification to minimize rework. In this way, the constraint would be making $2.50 per minute instead of $2.22. However, this decision must be made from a global perspective and should include the impact on other resources as well as the final product.

	(Scrap) Product B	(Rework) Product B
Throughput Generated	$ 75	$100
Total CCR Time per unit	30 min.	45 min.
Throughput per unit of Constraint Time	$2.50	$2.22

Changing the mean of a process may affect a product's acceptability at other processes. Products are designed so that the nominal value of each specification matches.

Those resources that are prerequisites to the constraint and have a C_pk of less than 1.33 should be judged, like all decisions, based on their impact on Throughput, Inventory and Operating Expense. Whether the correlation between SPC charts is positive or negative, the mean of the prerequisite resource should be moved to minimize the impact at the constraints' ability to create Throughput and the generation of Inventory and Operating Expense. Scrap at nonconstraint resources occurring before the constraint will diminish the value of raw material. Rework at those same resources has no value unless protective capacity is used.

Based on the product flow diagram, increasing capacity at resource R-1 would probably move the CCR to resource R-5, a resource that is loaded to 90%. If R-1 had a C_pk of 1.00 and R-5 had a C_pk of 1.33, this might be the right approach to take. However, there are other considerations. Moving the CCR to R-5 would also shorten the distance from the constraint to the sales order. Scrap at resources R-4 and R-1 would then play a much smaller role.

Once those resources that are critical to the generation of Throughput have been brought under control, the impact at other resources can be addressed. Those resources that create holes in the buffer caused by out-of-control conditions should be addressed as well. The buffer management process should indicate those resources that are causing consistent delays in reaching the buffer origin.

PRODUCTS DESIGNED AROUND THE MIDDLE OF THE SPECIFICATION

There is another consideration that has not been addressed: the cumulative impact of resources whose means are not grouped around the nominal value of the specification. Two parts that are designed to be fitted together but whose measurements are at the wrong ends of the specification may not fit at all. Nuts and bolts are a prime example of this. A bolt may have an outside dimension of ¼ inch ±.015. A nut may have an internal dimension of ¼ inch ±.015. The nominal value would be ¼ inch. However, if the mean of the process that produced the bolt is set at ¼ inch −.015, this would produce bolts that are undersized. If the mean of the process that produced the nut is set at ¼ inch −.015, this would be produce nuts that are undersized as well. The result would be mismatched parts. It may be easier to ensure that the means of processes are matched to the target value of the specification than to wait until there is an impact on Throughput.

THE IMPLEMENTATION SEQUENCE

As in the implementation of any complex process, a logical sequence of events should be used to assure a smooth transition. The following is a suggestion for the implementation of SPC.

1. Develop the product flow diagram
2. Establish buffer management
3. Identify critical operations
4. List operations in order of criticality based on their impact to Throughput
5. Choose the quality characteristics to be charted
6. Implement the pilot project
7. Schedule the remaining charts to be implemented
8. Organize a committee for each chart
9. Collect data, plot charts and develop C_pk indexes
10. Analyze charts as data becomes available
11. Correct assignable causes based on the impact to Throughput

In choosing the quality characteristic, it is very important to choose those characteristics that, if kept in control, will maximize Throughput. Candidates include those characteristics:

- That represent a large portion of Throughput generated by the constraint.
- Required by QFD and that are critical to the primary function of the product.
- That serve as prerequisites to the constraint.

- That may negatively effect work completed at the constraint.
- Whose rejection rate causes holes in the buffer.

In implementing a pilot project, it is wise to begin with a resource that is having a major impact on Throughput because of rework or scrap. Any success here will create the support needed to extend the process to other resources. If one is unaware of a specific resource that is having this type of problem, begin with the constraint. Once the resource is selected, begin the charting process and develop the $C_p k$ index. The remaining charts should be implemented based on their impact on Throughput, Inventory and Operating Expense.

In organizing a committee for each chart, the objective is to ensure that when an improvement must be found, the right people are there to find it. This committee may include a representative from engineering, the operator, the supervisor of the area and quality assurance. Each committee should have people who understand general problem-solving techniques such as brainstorming, fishbone analysis, design of experiments, and so on.

Buffer management can begin at any time during the implementation of SPC. However, it should be one of the first steps in the process so that the benefits of using the aggregated impact of problems on the buffer origin can be gained.

OVERCOMING BLOCKING ACTIONS

One of the biggest problems in making SPC successful is a lack of understanding of *what* it is and *where* it should be used. What SPC is can easily be solved through education. But where it should be used is a different issue. As with implementing TOC, many problems must be overcome. The basic blockers to a successful implementation include

- Conflicting goals and measurements
- Poor decision systems and support mechanisms
- Lack of understanding of how to meet the necessary conditions
- Lack of understanding of how resources interface and their resulting impact on the system
- The cost mentality
- Failure to understand how to motivate people

People may block the implementation of SPC in different ways: managers and supervisors, by issuing invalid priorities or withholding resources; workers, by charting incorrectly, refusing to chart or not reacting correctly to problems encountered. Most people can understand how to implement, the importance of SPC, where to chart, and so forth. But motivating managers, supervisors and workers to implement successfully motivation may be very difficult. Even though more company presidents are realizing that customers are making SPC a necessary condition by requiring SPC charts before placing order a successful implementation of SPC depends on the impact on profitability not just from the customer's perspective but from all aspects. Fulfilling the necessary condition ensures only that customers are happy; it does not ensure that profitability will continually rise.

The biggest problem in motivation are fear and "not invented here." (In theory the inventor is the only person not affected by fear or "not invented here.") These issues must be overcome. One method being used with growing success is the Socratic method, which supports the process of having people invent their own solutions. The "student" is placed in a situation, given a set of guidelines and then asked to find a solution. This can be done using simulators, by asking just the right question or by using the policy analysis/TOC thinking process described in Chapter 4. It is very important that the student not be given the answers, as it is the process of self-discovery that helps the student realize the truth.

STUDY QUESTIONS

1. What additional tools are needed to focus the implementation of SPC so that it is in line with the goal of the company?
2. What impact, in terms of where to focus, has TOC had on the implementation of SPC?
3. What are the most vulnerable portions of the factory with respect to controlling variation?
4. In what way does scrap or rework impact product mix?
5. What steps are needed for a TOC implementation plan?
6. What quality characteristics should be kept in control to maximize throughput?
7. List those actions that would cause a successful implementation of SPC to fail?
8. What is the Socratic method, and how can it be used to support the implementation of SPC?

12

TOC and Experimental Design

This chapter will use the Taguchi method as an introduction to the DOE process and will explore how the DOE process should be used in support of the five-step improvement process.

OBJECTIVES

- To create a basic understanding of the experimental design concept
- To design a two-level and multilevel experiment
- To conduct an analysis of variance through the use of the ANOVA table
- To gain insight into how to use DOE methods to support the exploitation and subordination processes
- To explore the validity of the Taguchi loss function as a focusing mechanism

DOE DEFINED

The objective of the designed experiment is to illustrate the impact of specific changes on the inputs of a process, and then to maximize, minimize or nominalize the outcome by manipulating the input. The designed experiment is usually used when it is unclear what impact a specific set of inputs may have, either individually or collectively, on a process. By constructing the experiment in a specific way and recording the resulting impact or observation, it is possible to understand what impact each variable within the process has, and the degree of result.

Multiple runs or trials are made while manipulating each input or group of inputs and the resulting observations are recorded. An analysis of the inputs as well as the result determines the level of input needed to arrive at the optimal result.

<u>Inputs</u>   <u>Output</u>

Man
_____

Material
_____

Machine
_____   ┌─────────────────┐
Method   │   │
_____   │   **Process**   │   **Result**
Tool   │   │
_____   └─────────────────┘
Environment
_____

DESIGNING THE EXPERIMENT

The first step in conducting the designed experiment is to determine the quality characteristic to be examined and the desired effect. This is probably the most important step and should be examined carefully. Brainstorming sessions are usually used to gain insight into what the goal of the experiment should be and to understand what factors (man, method, machine, material, tools and the environment), as well as levels within factors, might impact the goal. The fishbone chart is usually used to gather and organize information in brainstorming sessions of about 10–12 people with various technical backgrounds so that a thorough cause-and-effect investigation can be performed.

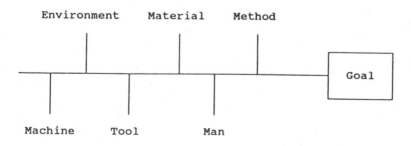

A level is a certain aspect of a factor chosen for the experiment. As an example, specific temperatures would be levels within the methods factor while drill bit types would be levels within the tools factor.

Factors	Tool	Method	Material	Machine
Levels	Carbide	100 Degrees C	Stainless	Lathe 1
	Titanium	150 Degrees C	Tungsten	Lathe 2

Once a set of factors/levels has been selected, specific treatments are determined. A treatment is a specific set of factors/levels combined into one test. Each test of the same treatment is called a trial. Each trial has an outcome or observation.

TWO-LEVEL EXPERIMENTS (THE TAGUCHI METHOD)

The two-level experiment is so named because it involves an experiment with a limited the number of levels (two within each factor). Once the goal for the experiment has been determined and levels to be tested have been selected, specific combinations of levels are arranged so that each treatment will be different. Then the test trials are run. The results are recorded on the orthogonal array. An orthogonal array is a matrix used to plan and control the experiment. The orthogonal array in Figure 12.1 presents four different treatments or mixtures of factor levels and two levels for each of three factors. Each treatment is read from left to right. Two trials or runs are held for each treatment and the results are recorded under observations.

Notice that for each treatment the mixture of levels is never the same. In treatment 1, all level-1 factors are used. In treatment 2, level 1 for factor 1 is mixed with level 2 for factors 2 and 3. In this way, each level is used with every other level. Notice also that each level is used twice within each factor. Level 1 for factor 1 is used in treatments 1 and 2. Level 2 for factor 1 is used in treatments 3 and 4 and so on. By designing the experiment so that one can determine the

Treatments	Factor 1	Factor 2	Factor 3	Observations	
1	Level 1	Level 1	Level 1	Trial 1	Trial 2
2	Level 1	Level 2	Level 2	Trial 1	Trial 2
3	Level 2	Level 1	Level 2	Trial 1	Trial 2
4	Level 2	Level 2	Level 1	Trial 1	Trial 2

Figure 12.1   The orthogonal array.

impact each level will have on the observations, it is possible to arrive at the optimal mix of levels that will produce the desired result.

For this experiment, which studies the problem of bridging encountered at the wave solder machine of a printed circuit board manufacturer, the factors of process speed, temperature and wave height are chosen. The objective is to minimize bridging resulting from the wave solder by finding just the right combination of speed, temperature and wave solder height. Specific levels for elements were chosen as follows:

	Low	High
Temperature	475 Degrees F	510 Degrees F
Wave Height	.25 inches	.30 Inches
Process Speed	4 feet per minute	5 feet per minute

Under the Taguchi method, the low measurements are placed in the level-1 positions within the orthogonal array, while the high measurements are placed in the level-2 positions. After the trials have been run for each treatment, the observations are also added.

For the experiment, observations are recorded as the total number of bridges per subgroup sample of five boards. The trial is run for a specific group of five boards and then bridging is measured. The next trial is run with the next group of five boards. This continues until all trials and treatments have been completed and observations recorded.

Treatments	Temp.	Height	Speed	Observations	
1	475 F	.25 in	4 fpm	14	10
2	475 F	.30 in	5 fpm	30	21
3	510 F	.25 in	5 fpm	20	10
4	510 F	.30 in	4 fpm	37	20

To complete the array, the actual measurements for each level and the names for each factor are abbreviated so that the array resembles that shown at the top of the next page.

Level-1 measurements are replaced by a "1," level-2 measurements are replaced by a "2" and each factor is abbreviated, in this cases by using the first letter of the descriptive title. This makes it easier to focus on analyzing the observations.

Treatments	T	H	S	Observations	
1	1	1	1	14	10
2	1	2	2	30	21
3	2	1	2	20	10
4	2	2	1	37	20

The first step is to determine how much variation has occurred among all treatments. The total variation that has occurred is determined by adding the square of each observation. This is commonly referred to as the sum of squares.

$$\sum x_i^2 = 14^2 + 10^2 + 30^2 + 21^2 + 20^2 + 10^2 + 37^2 + 20^2$$

$$= 3906$$

The next step is to distribute this variation among a number of different types of variation that can occur, including

- *Grand mean effect*: The variation caused by the item being tested, in this case, the wave solder machine
- *Effect of factors*: The variation caused by the individual factors within the experiment
- *Repetitional error effect*: The variation caused by the number of trials that are performed

EFFECT OF THE GRAND MEAN

The grand mean is determined by summing the observations, squaring them and then dividing by the number of observations

$$\frac{\left[\sum x_i\right]^2}{n} = \frac{\left[14 + 10 + 30 + 21 + 20 + 10 + 37 + 20\right]^2}{8}$$

$$= \frac{162^2}{8} = \frac{26244}{8} = 3280.5$$

EFFECT OF THE FACTORS

The effect of the factors is determined by adding together the level-1 observations for each trial by factor, adding together the level-2 observations for each trial by factor and then subtracting the level-2 factor from the level-1 factor and squaring the result for each factor.   The result is then divided by the total number of observations. The formula for the effect of temperature is written as follows:

$$S_T = \frac{\left[S_1 - S_2\right]^2}{n}$$

In Figure 12.2, the level-1 observations have been identified for temperature. Each observation for level 1 is added together.

$$14 + 10 + 30 + 21 = 75$$

In Figure 12.3, the level-2 observations have been identified for temperature. Each observation for level 2 is added together.

$$20 + 10 + 37 + 20 = 87$$

The effect of temperature is

Treatments	T	H	S	Observations	
1	1	1	1	14	10
2	1	2	2	30	21
3	2	1	2	20	10
4	2	2	1	37	20

Figure 12.2   Level-1 observations for temperature.

Treatments	T	H	S	Observations	
1	1	1	1	14	10
2	1	2	2	30	21
3	2	1	2	20	10
4	2	2	1	37	20

Figure 12.3 Level-2 observations for temperature.

$$S_T = \frac{\left[75 - 87 \right]^2}{8} = \frac{144}{8} = 18$$

In Figure 12.4, the level-1 observations have been identified for wave height. In Figure 12.5, the level-2 observations have been identified for wave height. The effect of wave height is

Treatments	T	H	S	Observations	
1	1	1	1	14	10
2	1	2	2	30	21
3	2	1	2	20	10
4	2	2	1	37	20

Figure 12.4 Level-1 observations for wave height.

Treatments	T	H	S	Observations	
1	1	1	1	14	10
2	1	2	2	30	21
3	2	1	2	20	10
4	2	2	1	37	20

Figure 12.5 Level-2 observations for wave height.

$$S_H = \frac{\left[51 - 108\right]^2}{8} = \frac{3294}{8} = \mathbf{406.13}$$

The effect of speed is

$$S_S = \frac{\left[84 - 81\right]^2}{8} = \frac{9}{8} = \mathbf{1.13}$$

REPETITIONAL ERROR EFFECT

The repetitional error effect can be determined by subtracting all other effects from the total variation. The formula for finding the repetitional error effect for the experiment is

$$S_{e2} = S_T - S_m - \left[S_T + S_W + S_S\right]$$

The repetitional error effect is

$$S_{e2} = 3906 - 3280.5 - \left[18 + 406.13 + 1.13\right] = \mathbf{200.24}$$

DETERMINING SIGNIFICANT FACTORS

The next step in the process is to determine which factors are important and which are not. While at this point it is obvious that wave height is far more important than speed and temperature, there are other things to consider. To understand the significance of each measurement, the minimum variance required for a measurement to be considered significant must be determined. This is calculated by dividing the repetitional error effect by the degrees of freedom and then multiplying the result times the F-value. The degrees of freedom are determined by adding 1 to the total number of factors in the experiment and subtracting the result from the total number of trials.

```
Degrees Of Freedom  =  Total Number Of Trials - Total Number
                       Of Factors  +  One

Degrees Of Freedom  =  8 - (3 + 1)  =  4
```

In this case, there are three factors and eight trials. Therefore, there are four degrees of freedom $[8 - (3 + 1) = 4]$. The F-factor is taken from the F distribution table in Appendix A.2. The degrees of freedom are used to find the F-value. In this case, the F-value is the intersection of row 4 and column 1 on the F-distribution chart and equals 7.71.

$$\text{Minimum Significant Variance} = \frac{\text{Repetitional error effect}}{\text{Degrees Of Freedom}} \times \text{F-Value}$$

$$= \frac{200.24}{4} \times 7.71 = \mathbf{385.96}$$

Variance For Wave Height $= 406.13$

Minimum Significant Variance $= 385.96$

Since the variance for wave height is greater than the minimum significant variance, wave height is a critical issue. Speed and temperature are considered nonsignificant. If a significant measurement is not found, the obviously insignificant factors can be dropped and the minimum significant variance recalculated. Since speed and temperature are not considered significant, the degrees of freedom need to be recalculated. In this case, there is only one factor in the formula

Degrees Of Freedom $=$ Total Number Of Trials $-$ (Total Number Of Factors $+$ One)

Degrees Of Freedom $= 8 - (1 + 1) = \mathbf{6}$

The F-value in this case would change (see Appendix A.2). Since the number of degrees of freedom has changed to 6, the F-value for a two-level experiment with one factor is 5.99.

$$\text{Minimum Significant Variance} = \frac{\text{Repetitional error effect}}{\text{Degrees Of Freedom}} \times \text{F-Value}$$

$$= \frac{200.24}{6} \times 5.99 = \mathbf{199.88}$$

DETERMINING OPTIMAL RESULTS

By determining which levels within the wave height factor create the best average results, the optimal setting for wave height can be determined.

Treatments	T	H	S	Observations	
1	1	1	1	14	10
2	1	2	2	30	21
3	2	1	2	20	10
4	2	2	1	37	20

$$\text{Average result (Level One)} = \frac{(14 + 10 + 20 + 10)}{4} = 13.5$$

The average result using the first level is 13.5.

Treatments	T	H	S	Observations	
1	1	1	1	14	10
2	1	2	2	30	21
3	2	1	2	20	10
4	2	2	1	37	20

$$\text{Average result (Level Two)} = \frac{(30 + 21 + 37 + 20)}{4} = 27$$

The average for the second level is 27. Since the objective is to minimize the amount of bridging on printed circuit boards, the best wave height for this experiment is level 1, or .25 in., while the settings for speed and temperature are insignificant.

CONDUCTING LARGER EXPERIMENTS

Experiments are often larger than the two levels, three factors and two trials presented in the preceding two-level experiment. To conduct larger experiments, a few basic changes are necessary; however, the concept is roughly the same. The key issue is to ensure that the experiment is constructed around the proper factors and levels before proceeding. In the wave solder experiment, if the shape of the wave becomes an issue, the number of trials increases from two to three, the number of factors increases and the number of observations also changes.

	Low	Med.	High
Temperature	475 F	490 F	510 F
Wave Height	.25 in.	.27 in.	.30 in.
Process Speed	4 fpm	4.5 fpm	5 fpm
Wave Shape	1	2	3

T	H	S	W	Observations			
1	1	1	1	15	27	22	17
1	2	2	2	30	21	36	27
1	3	3	3	20	25	22	10
2	1	2	3	5	0	13	10
2	2	3	1	44	29	37	42
2	3	1	2	30	27	30	21
3	1	3	2	20	15	17	25
3	2	1	3	17	29	20	24
3	3	2	1	37	28	37	20

Notice that the orthogonal array is much larger. The first step again is to determine how much variation has occurred among all treatments. The total variation is still determined by adding the square of each observation.

$$\text{Total Variation} = \sum_i x^2 = 23,443$$

The next step is to determine the effect of the grand mean as in the two-level experiment.

$$\text{The Grand Mean Sum of Squares} = \frac{\left[\sum_{i} X\right]^2}{n} = 20,022.25$$

Determining the effect of the factors for an experiment of this size is a bit different from for the two-level experiment. Each factor level observation is summed individually, squared and then divided by the number of observations at each level. The grand mean is then subtracted.

$$\text{Effect Of The Factors} = S_T = \frac{T_1^2 + T_2^2 + T_3^2}{n} - \text{Grand Mean Sum of Squares}$$

T_1 for this problem is 272 and is determined by adding all the level-1 observations for T at level 1, as highlighted below.

T	H	S	W	Observations			
1	1	1	1	15	27	22	17
1	2	2	2	30	21	36	27
1	3	3	3	20	25	22	10
2	1	2	3	5	0	13	10
2	2	3	1	44	29	37	42
2	3	1	2	30	27	30	21
3	1	3	2	20	15	17	25
3	2	1	3	17	29	20	24
3	3	2	1	37	28	37	20

T_2 is 288 and is determined by adding all the level-2 observations for T.

$$S_T = \frac{272^2 + 288^2 + 289^2}{12} - 20,022.25$$

$$S_T = \frac{73,984 + 82,944 + 83,521}{12} - 20,022.25 = \mathbf{15.16}$$

This process is repeated for each factor and level.

$$S_H = \frac{186^2 + 356^2 + 307^2}{12} - 20,022.25 = \mathbf{1276.16}$$

$$S_S = \frac{279^2 + 264^2 + 306^2}{12} - 20,022.25 = \mathbf{75.5}$$

$$S_W = \frac{355^2 + 299^2 + 195^2}{12} - 20022.25 = \mathbf{1098.66}$$

The repetitional error effect can be determined in the same manner as earlier. The formula for finding the repetitional error effect for the experiment is

$$S_{e2} = S_T - S_m - \left[S_T + S_H + S_S + S_W \right]$$

The repetitional error effect is

$$S_{e2} = 23,443 - 20,022.3 - \left[15.2 + 1276.2 + 75.5 + 1098.6 \right]$$

$$= \mathbf{955.2}$$

DETERMINING THE SIGNIFICANCE OF THE FACTORS

The analysis of variance (ANOVA) table is used to organize the analysis of variation that occurs during the designed experiment and to attribute variation to its source. Figure 12.6 is an ANOVA table for the preceding experiment.

The left-hand column indicates the source of the variance. The df column indicates the number of degrees of freedom assigned to each source. The degree of freedom for the mean is 1. The degree of freedom for each factor is the number of levels, that in this case is $3 - 1$. The degrees of freedom for the repetitional error effect (27) is equal to the total number of observations (36) minus the degrees of freedom assigned to all factors (8) and the grand mean effect (1). The S column represents the sums of squares and is the amount of variation found at each stage of the experiment, including the total variation, the grand mean, each of the factors and the repetitional error effect. The V column represents the average factor and error variance and is equal to the S column divided by the df column within source. The F-value is determined by dividing each factor variance under the V column by the value of the repetitional error effect, also found under the V column. To determine whether a factor has significance, the F-values are compared to the critical value found in the F-value table from Appendix A.2. The df for each factor (2) is located by moving across the top of the chart and the repetitional error (27) is located on the left side. The critical F-value is located at the intersection of the two. The critical value is 3.35. For the experiment, any F-value that equals or exceeds 3.35 is considered significant. The significant factors for the experiment are wave height and wave shape.

Source	df	S	V	F-Value
m	1	20,022.3	20,022.3	
T	2	15.2	7.6	.21
H	2	1,276.2	638.1	18.04
S	2	75.5	37.75	1.07
W	2	1,098.6	549.3	15.53
e 2	27	955.2	35.38	
Total	36	23,443.0		

Figure 12.6 ANOVA table.

ADDITIONAL CONSIDERATIONS

Additional considerations in designing the experiment include the interactions between factor levels and noise factors.   Interactions can occur between factors, so to maximize the overall effectiveness of the experiment, one factor level may be used in the presence of another. In the preceding experiment, it was discovered that wave height and shape were significant. However, a specific wave height may be required for a specific wave shape in order to gain the largest reduction in overall bridging. A test for interaction must be conducted.

Certain factors such as weather cannot be controlled but may have an impact on the success of the experiment. An attempt must be made to predict optimal factor levels under different conditions or noise levels that are not a controlled part of the experiment. The impact of noise must be determined.

DESIGNING THE EXPERIMENT TO INCREASE PROFIT

As stated earlier, the objective of the DOE process is to determine the impact of specific changes to the inputs of a process and then to let one maximize, minimize or nominalize the outcome by manipulating the input. From a global perspective, it should be clear at this point that an improvement made at a local level may have no impact or even a negative impact on profitability. The typical experiment designed to improve a quality characteristic at a specific resource may cause an increase in Operating Expense without creating an increase in Throughput. Experiments should be designed with this issue in mind.

• • •

During the initial design phase of an experiment, one should determine the objective and decide whether it should support the attainment of the necessary condition, the five-step improvement process, or both.

• • •

As in SPC, DOE must be studied based on how resources interface and their resulting global impact on the system. The product flow diagram and buffer management system help tremendously in identifying what needs to be improved and in analyzing the impact.

In addition, consideration should be given to the impact a change in factor levels will have on other quality characteristics not addressed by the experiment and their impact on the ability to properly exploit or subordinate. In the wave solder experiment, while temperature was not a factor in reducing bridging, it could have resulted in board warpage and major problems at other resources.

Consideration must also be given to the measurements desired in the outcome of the experiment. The nominalization of an outcome of an experiment may, for example, result in an equal amount of scrap or rework at a resource that appears between the constraint and the sales order and is loaded to 60%. The best alternative may not be the nominalization of the measurement if it results in an equal amount of scrap as well as rework. As seen in Chapter 11, for this portion of the net, a scrap will have far more impact than a rework.

PREDICTING THE IMPACT

Once the predicted outcome of the experiment has been determined, another prediction must be made for the impact of the experiment on Throughput, Inventory and Operating Expense. Determining the impact on Throughput from an overall reduction of processing time at the constraint can be made by understanding how a release of constraint time is to be used to create additional Throughput (see Chapter 9, on exploiting the constraint). This is best accomplished by looking at the overall schedule for the constraint. Key issues to consider include whether the experimental design is product specific or applies to all products created at the constraint. Remember that by changing constraint processing time, the Throughput per unit of the constraint also changes and may result in a change in the preferred product mix.

As stated earlier, determining the impact on Inventory and Operating Expense may be determined only after an attempt has been made to subordinate to the schedule created for the constraint in consideration of the predicted outcome of the change in factor levels (see Chapter 3). Once a change has been made that increases constraint availability, there will be an immediate impact on those resources that must react to the increase but that are themselves restricted in capacity.

SUPPORTING SPC

Design of experiments is also used to support the SPC program. When SPC has been used to identify an out-of-control condition at a resource that is causing a threat to exploitation or subordination, DOE can be used to improve at that resource and bring the process back into control or increase the $C_p k$ by manipulating the input to either maximize, minimize or nominalize the output.

PROBLEMS IN THE TAGUCHI LOSS FUNCTION

Developed by Genichi Taguchi, the Taguchi loss function equation is generally used to estimate the amount of loss associated with a specific measurement within

a process.   The amount of loss is estimated for a specific measurement and a
baseline established that is then used to estimate all other losses associated with
individual measurements for the same process.   The loss is computed using $L =$
$k(y - T)^2$ where k is a monetary constant, y is a response value and T is the target
value.

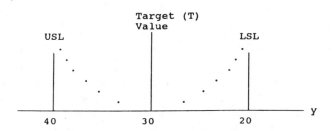

The value of k is determined by the following:

$$k \; = \; \frac{\text{Estimated Loss For y}}{(y \; - \; T)^2}$$

If the estimated loss for y = 20 equals $50, then

$$k \; = \; \frac{\$50}{(20 \; - \; 30)^2}$$

$$k \; = \; \frac{\$50}{100} \; = \; .5$$

The estimated loss for any measurement of y would be

$$L \; = \; .5 \; (y \; - \; T)^2$$

If y = 25 for the above, then

$$L = .5 (25 - 30)^2$$

$$L = .5 \times 25 = \$12.50$$

The Taguchi loss function assumes that k will be constant for all measurements of y and that the closer to the target value the lower the loss. This is an isolated view of what is happening, and in fact, there are many more issues to consider. Whether or not a monetary loss actually occurs depends on what resource is being considered, what its position is within the plant, the degree of resource load and what type of loss is being considered. The assumption that k will be constant for all measurements of y ignores the issue that on one side of the target value may be scrap and on the other may be rework. It also ignores the fact that one resource will have an impact on the measurements and monetary loss created at another resource. No one would argue that the further away from the target value, the more problems will occur. However, the degree of monetary impact is a different issue. The question that must be raised is, what is the impact on Throughput, Inventory and Operating Expense for any given measurement of y? As seen in Chapter 11 on implementing SPC, the impact on Throughput at the constraint depends on the amount of constraint time absorbed by the scrap or rework and the rate at that the scrap or rework occurs. The further away from the target value, the more constraint time is absorbed from replacing the scrapped part or accomplishing the rework. The impact on the nonconstraint may or may not have any monetary impact. As with SPC, to understand what is important, a method must be used to determine the relationships between resources so that global issues can be addressed. To do this, the product flow diagram is used (Figure 12.7).

Resource R-6 exists between the constraint and the sales order. Any given measurement of y can have a different impact on Throughput. If a given measurement of y were to cause scrap, then the loss would be the total value of the sales order, because the constraint must be used to replace the part. Therefore Throughput would decline. If a given measurement of y were to cause a rework, the additional capacity of R-6 could be used to accomplish the rework. Throughput would not decline. However, what would be the impact if, at a certain measurement of y, R-6, because of the increased rework, became constrained? If the constraint were large enough and surpassed the load at resource R-1, it would reduce the Throughput for the entire Throughput channel to whatever rate R-6 established. In Figure 12.7, R-6's load has been increased to 120% because of rework.

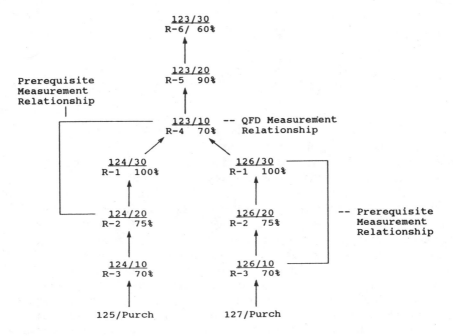

Figure 12.7 Product flow diagrams.

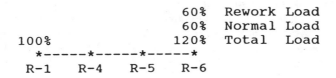

.   If R-6 were to have a rework load of 30% and not 60%, the impact would be a bit different.

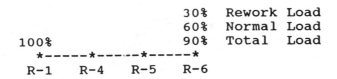

While R-6's load does not exceed R-1's, it will probably cause delays in shipment and an increase in inventory in front of resource R-6. Throughput would decline slightly because of the late orders, or operating expense would go up because of an increase in overtime to prevent late orders.

The measurement of one resource may have a direct relation to other resources within the net. Notice that R-3 has a direct relationship to R-1 through part operation 126/10. A given measurement of y causing a rework at 126/10, where excess and protective capacity would offset the impact could have a very negative effect on 126/30, where any loss of time would cause an immediate impact on Throughput.

In reality, the financial impact due to scrap or rework at a specific resource may resemble the lines A-B and B-C, where A-B is the loss in inventory and operating expense as the measurement begins absorbing protective capacity. The line B-C represents the creation of the primary constraint.

Notice that in Figure 12.9 that lines A-B and B-C have been shifted to the right but the loss remains exactly the same.

With the exception of the additional raw material required for replacing scrapped material, unless there is a decrease in Throughput or an increase in Inventory and Operating Expense because of the occurrence of scrap or rework, no change will occur due to a specific measurement on the y axis. This depends on the amount of protective, productive or excess capacity used.

To understand the global impact of a specific measurement of y, the following must be known:

- The location within the product flow diagram of the resource at which y occurred
- The location of the primary constraint
- An estimate of the volume of scrap or rework that will occur as a result of the measurement
- An estimate of the impact the additional resource load generated by the scrap or rework will have on nonconstraining resources
- The per-part cost of raw material, if scrapped

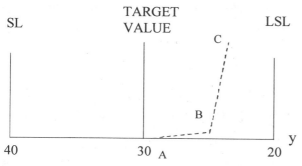

Figure 12.8 The impact of scrap or rework on a given resource.

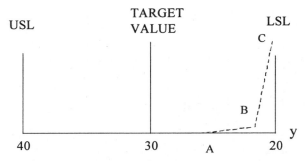

Figure 12.9 The impact of rework at a given resource.

- The per-product amount of sales involved
- An estimate of the impact of the measurement on other resources (prerequisite measurements)

Before any estimate of loss can be determined, an analysis of the physical environment must be made.

ADDITIONAL CONSIDERATIONS

Additional considerations include the loss due to additional wear by not producing products at the center of the specification. There is no argument that products should not be made as close to specification as possible. However, a monetary loss cannot be determined from the Taguchi loss function, which deals with local measurements and so cannot be used to judge the global impact with any certainty. Applying the Taguchi loss function may even result in misplaced priorities. From a local perspective, a resource may have a larger "cost" because of rework, but from a global perspective, the loss may not exist. The magnitude of a problem is not the only factor to be considered.

OTHER DOE TECHNIQUES

The Taguchi method has been used to introduce the DOE concept because it is easy to explain and easily understood. Other techniques exist as well. Each has good and bad points, but all are subject to laws governing the dependent variable environment as are the statistical methods presented here. For further study into DOE and other statistical methods, refer to Kiemele et al. (1991), Ryan (1989) and Doty (1991).

STUDY QUESTIONS

1. What is the objective of the design of experiment (DOE).
2. What is the first step in the designed experiment?
3. Define the terms *factor levels* and *treatments*.
4. What a the two-level experiment and from what is the name derived?
5. What is an orthogonal array and how is it used?
6. Define the following terms and explain how each is determined: grand mean effect, effect of the factors, repetitional error effect and degrees of freedom
7. What is the process for determining the significance of a given factor?
8. How are optimal results determined?
9. How is determining the effect of factors for a two-level experiment different from the determination for larger experiments?
10. What is the purpose of the ANOVA table and how is it used?
11. What is the objective of the Taguchi loss function and why does it fail?
12. What needs to be known to understand the global impact of a given measurement of y?

13

The TOC-Compatible Information System

OBJECTIVES

- To introduce the concept of the TOC-based information system
- To present methods and features for its use in the quality management environment

MANUFACTURING SYSTEMS TECHNOLOGY

Finding a physical constraint may be as easy as walking out onto the shop floor and looking for those resources where the Inventory is piled up or asking the expediters where they spend most of their time trying to get parts. However, for many companies it is not quite that easy. Trying to identify the constraint or constraints as well as the relationships between resources in a dynamic environment can be very confusing. Erroneous materials management strategies can cause situations where conditions of resource overload physically move from resource to resource without any logical explanation. Many technical paradigms in manufacturing systems technology such as rough-cut capacity planning, material requirements planning and capacity requirements planning make it virtually impossible to identify physical limitations. It is extremely important that these problems be overcome.

MANUFACTURING RESOURCE PLANNING

In most traditional systems, the approach to identifying physical limitations begins with the master production schedule (MPS). In creating the MPS, an attempt is

238

made to assess the load on all resources at the same time as creating a basic schedule for material requirements planning (MRP). Input to the MPS may include sales orders as well as forecasts of end (finished) items or spare parts, while a load profile representing each product within the MPS is used to determine basic demand requirements for specific resources and time periods. The input to MRP in the form of end items and due dates is then used along with current on-hand and on-order information as well as bill-of-materials information to identify that material required to produce the demand from the MPS. MRP feeds its output into capacity requirements planning (CRP) in the form of planned releases and due dates so that a detailed capacity report can be produced from routing and resource information.

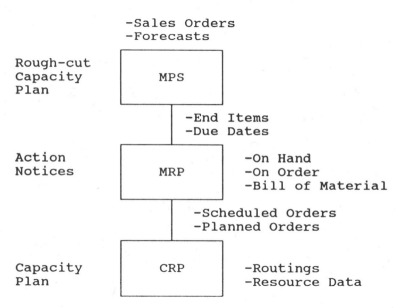

ADDRESSING THE PROBLEMS

There are three major problems that are fairly consistent throughout the traditional system and seem to violate basic laws governing the impact of product flow and the way in that resources interface. While these three issues do not represent all the problems, they are significant.

- Dynamic versus static data
- The aggregation of demand
- Interactions between resources

DYNAMIC VERSUS STATIC DATA

The first problem is that data that should be dynamic in nature, such as lead times and lot sizes, is maintained in the item master file as static. The following example is a typical gross-to-net requirements generation matrix used in creating the material requirements plan. The objective is to illustrate the computations used in generating action notices, for ordering material and for releasing or rescheduling work orders and purchase orders, and for generating input for capacity requirements planning.

Part: AFG	1	2	3	4	5	6	7
Gross Requirements				10			
Scheduled Receipts							
Planned On-Hand							
Net Requirements				10			
Planned Receipts				10			
Planned Releases		10					

Lead Time = 2

Ten AFG parts are required for gross requirements in period 4 as indicated by output from the MPS. Since there are none on hand or on order net requirements are also generated for period 4. The net requirement is a signal that a new order must be placed and a planned order generated for period 4. This in turn generates a planned release in period 2.

The question is, what determined the lead time for the planned release of 10 parts? Lead times are a function of the extent of "Murphy's Law," of the load that exists at those resources that are at or near full capacity and the size of the transfer batch. So, what produced this arbitrary 2-period lead time? It was generated as an estimate and placed in a file called the item master so that it could be used during the gross-to-net requirements process. The problem is it has nothing to do with what is happening on the shop floor at the time of the capacity estimation. It serves only to inflate inventories that block the creation of Throughput and make capacity planning impossible. This is a major problem recognized by JIT. The traditional solution is to lower the number of levels in the bill of material so that lead times are not exaggerated. This means that all material within certain levels are due at the same time, making it impossible to plan capacity.

THE AGGREGATION OF DEMAND

The second problem is that resource demand is aggregated within time blocks. Capacity is assumed available when, in fact, it may not be. The traditional method of finding a resource capacity that is constrained is to divide capacity by demand for a given horizon. In this way, demand is aggregated over the entire period and is matched to all capacity for the same period. Capacity is considered available at the time an order is to be processed when, in fact, it may not be.

A capacity report may graphically resemble Figure 13.1, where even though the capacity is well below the demand for a given period, a capacity limitation could still exist.

In reality, demand for any given day within the schedule may have large peaks and valleys where at one moment there is no demand, and at the next moment, demand may be twice the available capacity (Figure 13.2).

A certain amount of time is required to protect the scheduled delivery of each order. To prevent using the protective time and invalidating the sales order, excess demand must be scheduled earlier not later. Any capacity that exists after the excess demand must be scheduled, in this case to the left, and not aggregated across the entire period.

To determine whether a capacity limitation exists, that the cumulative effect of excess demand must be known for the entire period. To do this, a schedule must be created that begins at the end of the horizon and schedules excess demand

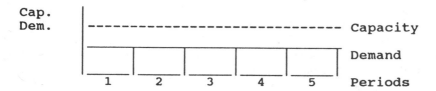

Figure 13.1 Traditional capacity measurement.

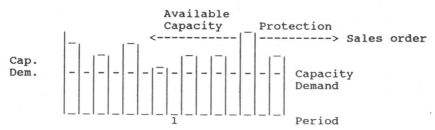

Figure 13.2 The impact of reality on capacity measurement.

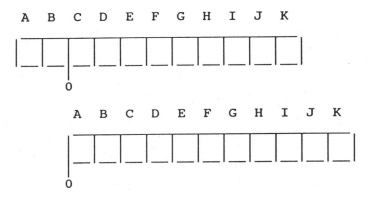

Figure 13.3   Nontraditional adjusting for time zero.

constantly toward time zero, allowing for enough protection to ensure that the sales order can be made on time.   There are two issues: (1) the resource that has managed to push the most demand into an earlier time period and (2) those orders that will appear past time zero.   That resource that has the highest amount of load pushed into an earlier time period, allowing for adequate protection to meet the sales order due date and has resulted in pushing the load the farthest past time zero, is the prime candidate for the constraint.

    To form a valid schedule means pushing the demand forward into the present at the primary constraint and then determining whether protective capacity is available on the non-constraints so that this new schedule will not be threatened (Figure 13.3).

    In dealing with the nonconstraints, the method and problems are somewhat similar, except that demand for all other resources must be determined by the schedule created at the primary constraint and not the sales order.   Whenever it is discovered that additional resources suffer from the same cumulative effect of a lack of adequate protection, it may be necessary to declare them as secondary constraints and create a schedule created for them as well, ensuring that the time for each is maximized.

INTERACTIONS AMONG RESOURCES

The third problem is that because each resource within a chain of resources has a direct impact on the schedule produced for the next, constraints can be located and dealt with only one at a time.

```
          A     B
     *----*----*----*
          90%   100%
```

Resource A is a near-constraint loaded at 90%. Resource B is the constraint loaded at 100%. In order to exploit resource B so that more Throughput can be generated, a schedule that maximizes B's time is generated, increasing B's load.

During the scheduling process, it may be found that by combining certain orders more Throughput can be generated. Until the outcome of the combination process is known, no schedule or load can be determined for any other resource within the chain, nor is it possible to determine the sequence to meet B's demand.

THE ROLE OF THE TOC-COMPATIBLE INFORMATION SYSTEM

In small, medium and large companies, depending on the complexity and nature of the business, it may be desirable to use a computer-based information system as an analysis tool for environments where resources are dependent on each other and are subject to variation in their ability to produce. The TOC-compatible information system is designed to react to the nuances of the dependent-variable environment through simulation. It clarifies the relationships among resources and acts to simulate the impact of external effects on the current balance or lack of balance among resources. The objective is to present a platform for analyzing and focusing the improvement program under the TOC umbrella. It should be capable of helping to answer questions such as what the impact of a certain setup reduction program on the other resources might be at a specific resource on the other resources in the plant and, ultimately, how this would affect return on investment.

In addition, the TOC-compatible information system should also lend insight by showing the *absence* of a particular influence. As an example, excess capacity is an indication that a certain marketing constraint exists. The absence of demand at resources that have more than enough capacity to create additional product at the cost of raw material is an indication of a poor market segmentation policy. While this may not be the case, it should lead to a study of how to exploit the excess capacity and help to provide a road map for the development of new products or the phasing out of old ones.

The TOC-compatible information system is vital in being able to

- Identify physical constraints
- Identify policy constraints

- Initiate valid schedules
- Implement improvement programs

APPLICATIONS

Engineering is able to use TOC-compatible information systems to support a wide variety of activities including total productive maintenance; setup reduction; statistical programs such as SPC, DOE and multivariate analysis; design for manufacturability, and process improvements. Finance is able to validate quickly any requests for capital equipment by Throughput justification.

Perhaps the most important application is in analyzing the strategic positioning of the company. It is an integral part of the recession-proofing process helping to determine what products should be sold in what markets, what the economic impact is of buying an additional plant and how the company should grow to minimize the impact of a downturn in the economy.

CONSTRUCTION

The TOC-compatible information system works by creating a valid schedule, since this is what ultimately determines the capabilities and limitations of the system. To do this, it must "understand" the way in that resources interface and react accordingly. When the schedule is produced, one must know the demand for each resource, the extent of excess and protective capacity and the relationships among resources. Any change in the environment affecting the schedule will have a direct impact on Throughput, since Throughput is created by the delivery of the product. Inventory and Operating Expense exist to support the schedule and are affected by either increasing or decreasing demand created by changes in the schedule. The model shows how Throughput, Inventory and Operating Expense are affected by changes in the schedule.

SUPPORTING THE DECISION PROCESS FOR QUALITY ASSURANCE

The constraint should be fully utilized in generating Throughtput. From a quality perspective, this means that

- Parts processed by the constraint should not require rework
- After being processed at the constraint parts should not be scrapped

Minimizing Inventory and Operating Expense requires reducing the problems that prevent the reduction of protective capacity (those things that create holes in zone 1 of the buffer).

In addition to identifying the constraint and providing the buffer management process, the system should supply an idea of how resources interface, in other words, which parts/operations and resources

- Feed the constraint
- Diverge
- Converge
- Appear after the constraint

Trying to produce product flow diagrams to define these issues can be a tedious effort. Often the first set of diagrams requires changing as soon as the ink is dry. Two visual tools supplied by the information system can help tremendously in this effort. These are the *where-used* and the *descending parts/ operations* screens.

THE WHERE-USED AND THE DESCENDING PART/OPERATIONS SCREENS

These screens are actually visual representations of all the product flow diagrams used to produce the current schedule (the net). The descending parts/ operations screen starts at the sales order and presents all part operations and the resources at which a specific sales order is to be processed. The where-used screen starts at the raw material and traces upward through all part operations and resources to all sales orders that may be using the part (Figure 13.4).

Notice that the constraint is highlighted. If there is a problem that exists at a resource prior to or after the constraint such as a capability index (C_pk) of 1.00 at resource R-5, part/operation B/20, it is easy to see that B/20 feeds the constraint and that a low C_pk could cause some serious problems for the constraint's ability to perform. It is also easy to see that if any scrap occurs on part/operation B/30 through A/30, the loss would be the price of the entire sales order. In addition, if there is a problem in receiving a particular purchase order on time or if there is a quality problem that affects raw material part C, the where-used can identify the parts/operations and sales orders that are impacted.

An abbreviated version of these screens includes those parts/operations and resources that appear after the constraint and feed the sales order (the "red lane") as well as those parts/operations and resources that feed the constraint (Figure 13.5).

Descending	Where used
S/O111 - A	RM C
A/30 R-1	B/10 R-6
A/20 R-2	B/20 R-5
A/10 R-3	B/30 **R-4**
B/30 **R-4**	A/10 R-3
B/20 R-5	A/20 R-2
B/10 R-6	A/30 R-1
RM C	S/O111 - A

Figure 13.4 Where-used and descending parts/operations screens.

Red Lane	Feeding the Constraint
S/O111 - A	B/30 **R-4**
A/30 R-1	B/20 R-5
A/20 R-2	B/10 R-6
A/10 R-3	RM C
B/30 **R-4**	

Figure 13.5 The red lane and resources feeding the constraint.

A modification of the descending parts/operations and where-used screens includes the addition of some important indicators (Figure 13.6).

Notice that rework and downtime information appears only for resources R-2 and R-4. To place this information on the system would only cause confusion if there is additional capacity available, there is no need to report the information. Additionally, if there were a problem at these resources, it would appear in buffer management anyway. The objective of this screen is to point the activity toward those resources that have critical capacity shortages. Scrap appears only for those resources that go from the constraint to the end of the process (the red lane).

These are important keys in focusing the quality program. But additional information is still needed, such as at what parts/operations the scrap or rework is occurring (Figure 13.7).

Parts/operations I/30 and A/30 are absorbing all of the scrap or rework time.

Throughput Chain 1

PRODUCT A	PRODUCT B		
PRODUCT C			
Resource	Scrap	Rework	Down Time
R-1	.01		
S R-2	.05	.10	.07
R-3	.00		
P R-4	.02	.03	.05
R-5			
R-6			

Figure 13.6 Key indicators.

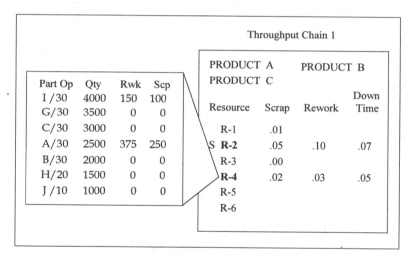

Throughput Chain 1

Part Op	Qty	Rwk	Scp
I /30	4000	150	100
G/30	3500	0	0
C/30	3000	0	0
A/30	2500	375	250
B/30	2000	0	0
H/20	1500	0	0
J /10	1000	0	0

PRODUCT A	PRODUCT B		
PRODUCT C			
Resource	Scrap	Rework	Down Time
R-1	.01		
S R-2	.05	.10	.07
R-3	.00		
R-4	.02	.03	.05
R-5			
R-6			

Figure 13.7 Key indicators (detail).

MAKING THE SCRAP/REWORK DECISION

A decision that involves both production and the quality organization is knowing when to scrap or rework a part. As seen earlier, quality cost will not work. Figures 13.8–13.10 represent the data required to make the decision and can serve as a basis for what features might be required for the information system to help in this process.

Product A takes 15 minutes to process on the CCR and uses $25 in raw material for each part produced. An order of 100 is being processed at a sales price of $100 each. After 50% of the order has been processed, 10 parts are

PRODUCT	CCR TIME/UNIT	CCR TIME AVAIL
A	15	1500 MIN
ORDER = 100	SP = $100	RM. = $25
DEFECTIVE PARTS	TIME/UNIT	MATERIAL
10	10	$5

Figure 13.8 The scrap/rework decision.

			(T)	CCR TIME
THROUGHPUT	=	60 (100 - 25)	= $4,500	900
MATERIAL	=	10 X 25	= - $ 250	
			$4,250	900

$$T/uc = \frac{\$4,250}{900} = \$4.72$$

Figure 13.9 Revisiting the foreman's decision.

				(T)	CCR TIME
THROUGHPUT	=	10 (100 - 30)	=	$700	100
THROUGHPUT	=	50 (100 - 25)	=	$3,750	750
				$4,450	850

$$T/uc = \frac{\$4,450}{850} = \$5.23$$

Figure 13.10   Revisiting the foreman's decision.

found to be defective. It will take an additional $5 in raw material and 10 minutes of constraint time to repair each one. The question is should the 10 parts be scrapped or reworked?

If the 10 parts are to be scrapped, then they will need to be replaced to fill the order. 60 additional parts are to be processed, creating Throughput of $75 (100 − 25) each, for a total Throughput generated of $4500. Since each part takes 15 minutes of constraint time and there are 60 parts, the total constraint time consumed will be 900 minutes. To replace the raw material scrapped for each of the 10 parts, $250 must be subtracted from the total Throughput generated. To scrap and replace the 10 parts and process the remaining order will take 900 minutes and generate $4250, or $4.72 per minute of constraint time.

To rework the 10 parts requires that an additional $5 raw material be subtracted from the sales price, which will generate $70 (100 − 30) of Throughput for each of the 10 parts processed, for a total Throughput of $700. It will take 100 minutes of constraint time to process all 10 parts. The remaining 50 parts will take 750 minutes of constraint time and generate $3750, for a total of $4450 generated with 850 minutes of constraint time, or $5.23 per minute.

The Throughput per minute of the constraint for the rework is greater, so the better choice would be to rework the parts.

The decision can actually be made from much less data than given. Once the parts causing the problem have been processed at the constraint, the question becomes how to maximize the Throughput of the system, given the current situation. The first issue is that the parts must be replaced so that the sales

			(T)	CCR TIME
THROUGHPUT	=	10 (100 - 5)	=	$950	100
THROUGHPUT	=	10 (100 - 25)	=	$750	150

	Rework				Scrap		
T/uc	=	$\dfrac{\$950}{100}$	=	**$9.50**	T/uc =	$\dfrac{\$750}{150}$	= **$5.00**

Figure 13.11 Revisiting the scrap/rework decision.

order can be delivered. At first glance, it will take $5.00 of additional raw material and 10 minutes of the constraint time to generate $100 for each part if the parts are reworked. If they are scrapped, it will take an additional $25 of raw material and 15 minutes of constraint time to make the $100. Throughput per unit of the constraint for rework is $9.50. Throughput per unit of the constraint for scrap will be $5.00.

As shown in Figure 13.11, the only data necessary to make the decision is

- The original cost of raw material for the product
- The sales price of the order
- An estimate of the amount of additional raw material and constraint time required to repair
- The quantity of the parts to be reworked or scrapped
- The original constraint time for the part

				T/uc	
Part/Operation	A/30		Scrap:	$5.00	
			Rework:	$9.50	
CCR Time Est. (Rework)	10				
			Product:	A	
Add. Raw material (Rework)	$5.00		Sales Order:	S/O111	

Figure 13.12 Providing information for making the decision.

The question now is what can the system do to provide this kind of information? Figure 13.12 represents an interactive screen that can be created to help with the scrap/rework decision. The only data needed to be entered is the part/operation being considered, the estimated constraint time to repair and the additional raw material required. Additional data such as the sales price of the order, the original constraint time and the original raw material cost can be obtained from the manufacturing system data base.

THE IMPACT OF SCRAP AND REWORK ON THE SCHEDULING PROCESS

Whenever a scrap or rework occurs, there is an impact on the amount of labor and material required by the system. This must be reflected in the scheduling process. Figure 13.13 is a product flow diagram for product A, which has a sales order demand for 10 units. The numbers appearing between each part/operation are the bill-of-materials relationships. Notice that for A/30 to be processed, it must receive one A/20 from resource R-2. Notice also that resource R-2 scraps 25% of the A/20s it creates.

This means that for every A/30 created, 1.33 units of A/20 must be processed first. To process 1.33 units of A/20, R-2 must receive 1.33 units of A/10. To produce all 10 units of A for the sales order will require 13.3 units of raw material C.

Scrap affects not only material requirements but also resource requirements. In order to obtain 1 unit of A/20 to be transferred to A/30, resources R-2 and R-3 must process the additional material. Additional processing will obviously take more resource time.

This kind of processing must be considered when creating the net and determining the load. It is done by modifying the bill-of-materials relationship during processing.

Rework is considered somewhat differently from scrap. Scrap is simply yielded during processing. However, rework will usually require a modification of the net (Figure 13.14).

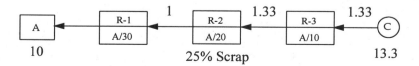

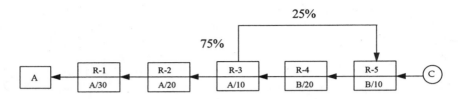

Figure 13.14 The impact of rework on the scheduling process.

Notice that 25% of all A/10s are reworked back to the beginning of the process. This means a certain increase in the load for resources R-3, R-4 and R-5. It also means, provided a constraint schedule is involved, that the rods will have to be adapted to ensure that reworked material maintains a specific distance in time from the process time of the original parts.

CONCLUSION

The manufacturing information system is undergoing tremendous change. In the quality sciences, this change is most apparent in the impact of the TOC-compatible information system. For a complete discussion of the TOC-compatible information system, see Stein (1992).

STUDY QUESTIONS

1. What is the purpose of the TOC-compatible information system? Upon what stepped process is it based?
2. What problems must be overcome for the traditional MRP system to be effective in implementing a quality program?
3. What is meant by the term *aggregation of demand*, and what is its significance?
4. How does the TOC-compatible information system re-create the relationships between resource?
5. What is the role of the TOC-compatible information system?
6. What is the benefit provided by the *where-used* and *descending part* operations screens? How are they constructed?
7. Make a scrap/rework decision from the data on p. 253.
8. What data presented in Question 7 was necessary to make the decision, and how can the TOC-based information system provide it?

PRODUCT	CCR TIME/UNIT	CCR TIME AVAIL
B	20	2400 MIN

ORDER = 150	SP = $200	RM. = $75

DEFECTIVE PARTS	REWORK TIME/UNIT	REWORK MATERIAL
25	15	$5

14

TOC in Supply Chain Management

This chapter emphasizes the importance of supply chain management and its overall impact on a dynamically competitive market. It expands the concept and use of TOC to a wider, more complex system comprised of chains of companies.

OBJECTIVES

- To establish the importance of using TOC throughout the entire supply chain
- To present a methodology for supporting such an effort

Managing the supply of parts coming into a company should begin long before raw material reaches the back dock or receiving area. However, the extent that this kind of management is needed far exceeds the expectations of most companies. The three very strong reasons to implement supply chain management arise from a company's need to

- Reduce Operating Expense
- Deliver on time
- Support product evolution

In today's competitive environment, there is tremendous pressure to lower prices. As prices are lowered, there is an expectation that profits must remain the same or increase. This has resulted in a tremendous amount of pressure to reduce Operating Expenses. The need to reduce Operating Expense implies a need to decrease inventories. Decreasing inventories have an important impact for on-time delivery. To a large number of vendors, reduced customer inventories has simply meant increased inventories for the vendor.

Exacerbating the situation is the rate of engineering change. Customers are demanding and receiving more new product innovations. This means that, as products change, vendors holding large amounts of Inventory risk holding raw material for customers who no longer need the parts. In situations where this is commonplace, an agreement can be made for the customer to buy the obsolete material from the vendor to ensure that the vendor will continue to agree to hold excess Inventory for them. But buying obsolete material from the vendor means an increase in Operating Expense for the customer. The dilemma is defined in the following evaporating cloud (assumption model) (APICS, 1996).

The objective written in the cloud is to (A) support building products that solve the markets demand for newer, more innovative products and to deliver on time. To do this, the vendor must (B) ensure that material will arrive to the customer on time to meet the customer's schedule and (C) respond quickly to frequent engineering changes. To ensure that material will arrive to the customer on time to meet the customer's schedule, the vendor must (D) maintain high inventories for the customer. To respond quickly to frequent engineering changes and to keep from having a large amount of chargebacks to the customer, the vendor must (D') maintain low inventories for the customer. The dilemma, of course, is that the vendor cannot hold both high and low inventories at the same time. Or, can this be done?

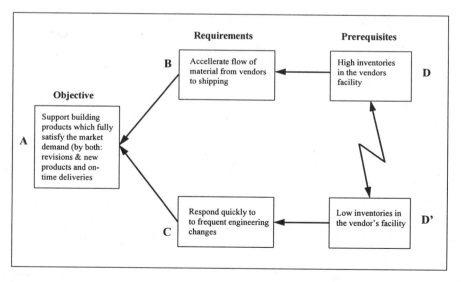

Figure 14.1. The supply chain dilemma.

Resource chains, regardless of size and complexity, always obey the physical laws that govern the dependent-variable environment. Whether the chain involves one company, a group of companies under the same corporate banner or individual disparate companies linked within the same supply chain, the chain will always respond in the same manner. Therefore, to affect supply chain management, the entire chain must be treated as one Throughput chain. This means that the five-step process of improvement must be implemented for the entire supply chain. The primary constraint must be identified and exploited, while the remaining resources are subordinated in the way in that it has been decided to exploit the constraint.

It is possible that the constraint for the chain may exist in one company and resources from another company must be used to subordinate to it. This makes the entire concept more difficult but obviously not impossible. If a drum-buffer-rope (DBR) schedule can be produced for the entire supply chain, even though it crosses corporate boundaries, resources will be activated for the right reasons, protection will be applied at the right locations and the Throughput generated by the primary constraint will be maximized. The result will be lower Inventory throughout the chain, better on-time performance and less writing off due to obsolescence. An additional benefit is that each participant in the project will create excess capacity. This capacity can be used in segmented markets to generate additional sales and, as seen in Chapter 5, when excess capacity is used, the difference between the sales price and the cost of raw material goes directly to the bottom line. This means that vendors can afford to reduce prices in order to meet customer demand.

DEFINING THE PROBLEM

Figure 14.2 represents three companies involved in a supply chain. Company A is being supplied by companies B and C. Notice that Inventory buffers have been placed prior to each resource and between the companies. This is the traditional model provided by not only MRP but also JIT. While JIT provides less Inventory within each buffer, the question that must be answered is where should Inventory be located in order to protect the Throughput generated by the entire Throughput chain?

The absurdity of this situation immediately surfaces when the capacity of each resource within the supply chain is known (Figure 14.3).

Notice that the capacity-constrained resource (CCR) is located in company B at resource 4. Notice also that there are no other resources within the chain that are loaded to near-constraint levels. There is no need for the amount of Inventory buffer applied to this supply chain. However, even if company A

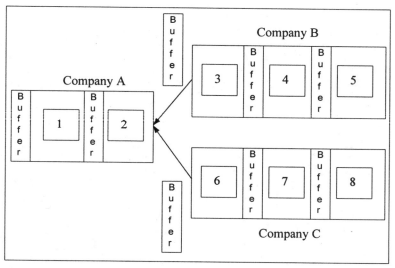

Figure 14.2 The supply chain.

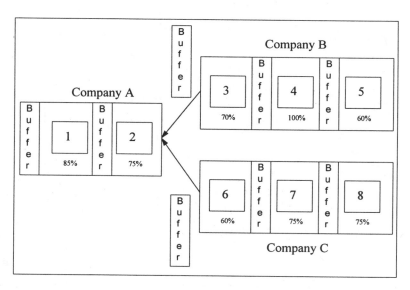

Figure 14.3 Adding capacity information to the supply chain.

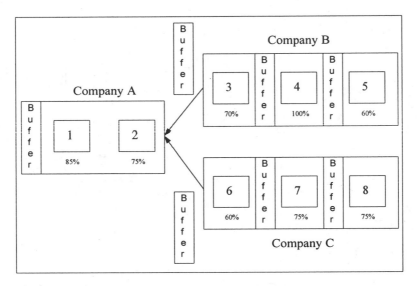

Figure 14.4   A partial solution to the situation shown in Figure 14.3.

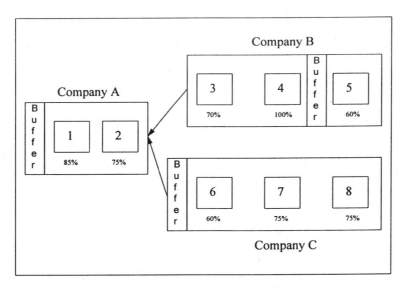

Figure 14.5   The buffer requirements for the supply chain.

introduces a DBR process into the plant and solves its Inventory problem, there is still the problem of the excess Inventory in companies B and C (Figure 14.4).

While the Inventory buffers have been placed correctly in company A, the excess Inventory in B and C will have a negative impact on company A.

From this perspective, it immediately becomes obvious that the entire supply chain should participate in the Inventory management process. As seen in Chapter 6, the place to buffer the chain of resources within one production facility is the constraint, shipping and the non-constraint legs for assembly operations that are fed by a constraint leg. This does not change, regardless of the length of the chain or the number of companies involved (Figure 14.5).

The place to buffer the system for these three companies acting as one Throughput chain under the present circumstances is the constraint (resource 4 of company B), shipping (for Company A), and the assembly buffer (resource 2 for company A).

This does not mean that company A should be satisfied with having a vendor with a capacity-constrained resource or that it should not try to pull the CCR into company A. However, unless company B actually knows that it has a capacity-constrained resource, it cannot begin to solve the problem of its customer. In addition, depending on the strategy decided upon by company A, it may be desirable to have the constraint in company B. It may be easier to control; it may even be easier to elevate as demand is dictated by the market.

The added benefit for company A is that the reassignment and lowering of Inventory buffers for companies B and C as well as the use of a more effective scheduling method enhances A's ability to deliver on time and has the potential of lowering the cost of raw material and obsolescence for A. The primary issue for this supply chain is one of subordination to the market of A. In generating the schedule for all resources within the chain, the first issue is how to deliver the sales orders for A on time. The resources of A as well as those for B and C must be subordinated to the schedule of the sales orders of A. The attempt to subordinate the supply chain to the sales order schedule for A will uncover the limited capacity at resource 4 in company B. To exploit resource 4, a schedule should be created and information fed back to company A so that the sales order schedule for A can be modified. (Understanding how to place the constraint discussed in the section "Placing the Constraint.")

THE IMPACT ON QUALITY

As with any chain, the string of resources leading from the constraint to the sales order is the most important. As discussed earlier, any scrap within this

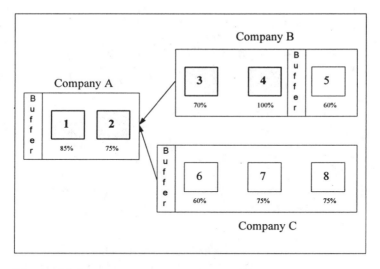

Figure 14.6 The red lane between companies.

"red lane" means an immediate loss in the full value of the sales order at company A. This is an interesting situation, since the profitability of A is directly attributable to the subordination of the quality efforts to prevent scrap at company B (Figure 14.6).

Since resources 1 and 2 depend on those measurements created at company B, it is vitally important that those prerequisite measurements needed to prevent scrap at company A be generated at company B. So the quality program of company B must also be subordinated to prevent any scrap after

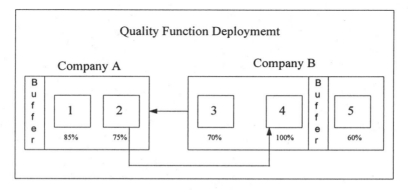

Figure 14.7 Quality function deployment.

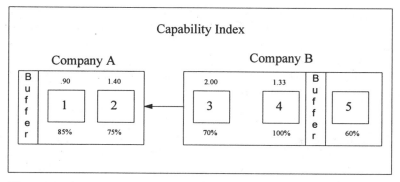

Figure 14.8   The capability index relationship between companies A and B.

the parts are delivered to A.   The quality function deployment process is very important here (Figure 14.7).

Notice that a prerequisite measurement has been established between resource 2 in company A and resource 4 in company B.   Additionally, if company A has a resource that is not capable due to a low C_pk as in resource 1, then no matter how good the quality delivered by company B, company A will still have a serious problem (Figure 14.8).

THE AGGREGATION OF PROTECTIVE BUFFERS

Aggregating the amount of protection necessary for company A to meet the demands of the market uses time buffers that must cross company boundaries (Figure 14.9).

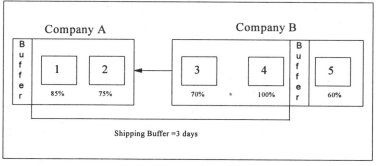

Figure 14.9   Time buffers crossing company boundaries.

Notice that the 3-day shipping buffer established for company A extends to resource 4 at company B. When the buffer manager for shipping at A is expediting those orders that have not arrived at shipping prior to zone 1, the manager must consider the situation at both companies A and B. Those delays that have caused the hole in zone 1 of the shipping buffer for company A are placed on the buffer manager's worksheet. If there is evidence that certain occurrences from company B are creating the majority of problems, the buffer manager for company A must notify company B. This will focus efforts on reducing those incidences in company B that threaten shipping at company A. By reducing these incidences, the amount of protection needed is also reduced. The shipping buffer at A can be shrunk, thereby lowering Inventory at companies A and B.

Notice the increase in proactive communication between the two companies. Company A is no longer expediting everything that shows up as being late on the purchase order list or expediting from company B only after something is late. The process of obtaining material from B is now subordinated to the needs of the shipping buffer for A. The need to share information on very specific quality/improvement issues becomes very clear. Buffer management helps to answer the questions of what and when to communicate between companies.

MARKET SEGMENTATION

As companies A, B and C begin generating excess capacity, each company will be able to segment the market so that resources can be protected and so that each can begin taking advantage of the new pricing flexibility they have just established. Since the difference between the sales price and the cost of raw material for those companies selling excess capacity is all profit, an additional advantage for vendors B and C is that as more profit is generated by each of the new market segments, more money becomes available to effectively deal with the demands of company A's customers for lower prices (Figure 14.10).

Notice that company C has acquired a new customer—company D. Notice also that an increase has occurred in demand for company C's resources but has not created another constraint. Product sold to company D is made with excess capacity.

If enough product is sold to company D that protective capacity is used, then company C will begin to threaten the schedule for company A by not being able to supply matching parts to resource 2 for those parts arriving from company B, resource 4 (the primary constraint). If a primary constraint is

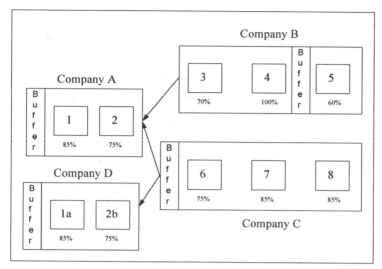

Figure 14.10 Adding a new customer.

generated within company C by the new sales, then for company A, resource 2 is being fed by two constraints (Figure 14.11).

Notice that resource 7 in company C is now loaded to 100%. Notice also that a new buffer has been established. An additional consideration for company C is an increase in Inventory and Operating Expense as a result of the addition of an internal constraint.

While company C may be successful in delivering to its new market on time (provided it is scheduled properly), the impact on A can be very different. Both companies B and C are limited in their ability to deliver on time. The probability that company A will receive matching parts from companies B and C is significantly effected. If resources 4 and 7 have no additional capacity to make up for those things that can go wrong and something happens to either resource, the supply of parts will be immediately out of balance. This will be a chronic problem. For company A to be successful, it must have additional capacity to protect itself from this situation. It must have enough capacity to allow for those things that go wrong at resources 4 and 7 in two other companies. With resource 1 in company A loaded to 85%, there may not be enough protective capacity for A to deliver on time. The more vendors company A uses with internal capacity constraints in the same Throughput chain, the lower the probability that A will be able to deliver on time. Without those matching parts, A cannot begin to process products for its customers on time and therefore will deliver consistently late.

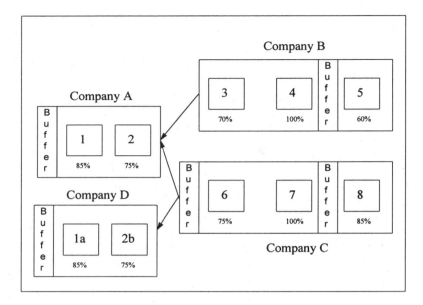

Figure 14.11 The impact of two primary constraints.

Company A does not realize what has happened. All it knows is that parts are not available from either company B or C to meet its schedule. Company A must now look for other sources for parts, while companies B and C lose a good customer. It is extremely important that the participants in a cooperative supply chain understand how and where to place the constraint and how to use excess capacity to generate profitability without endangering the effectiveness of the supply chain.

PLACING THE CONSTRAINT

To place the constraint within the supply chain correctly, some key issues are

- The ability to increase capacity incrementally so that the location of the constraint can be maintained/controlled.
- The stability of the technology applied to avoid scrap and material obsolescence.
- The versatility of the resource.

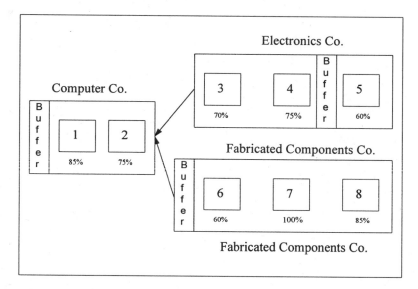

Figure 14.12 Placing the constraint.

- The amount of constraint time used relative to the creation of Throughput for the products being sold (Throughput per unit of the constraint).

In Figure 14.12, a computer manufacturer whose market is growing at a relatively rapid rate is being fed by two different companies—an electronics parts manufacturer and a fabricated components company.

Notice that the constraint is at resource 4 in the electronics company. Anyone who has been involved in the manufacture, sale or use of computers knows that electronics components are highly volatile. What passes for state of the art today will be obsolete in six months. Having the constraint in the electronics company raises several considerations:

- Because the technology is constantly changing, obsolescence for those parts that have already passed the constraint is a major problem
- Any scrap due to the infant mortality rate of electronics parts that have passed resource 4 will result in the loss of the entire sales price for the end item sold at the computer company
- The constant need to "keep up" with the latest technology may make it more difficult to maintain the constraint at a given resource

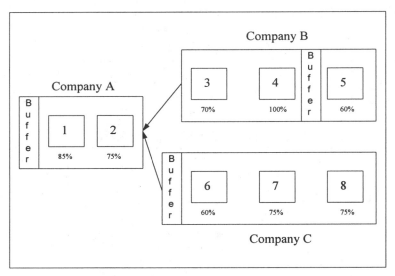

Figure 14.13 Placing the constraint.

- Manufacturing electronics components requires a higher degree of work-manship and expertise
- There is a higher degree of probability that an industry shortage, such as for integrated circuits, will occur

In this case, the weakest link in the supply chain would be associated with that company whose product is highly volatile with respect to engineering and technology changes. The possibility of scrap due to obsolescence is high, and the technology does not always work as well as expected. Depending on the methodology used, machines can be very costly to maintain and replace, making a gradual and controlled elevation of the constraint somewhat difficult. However, if the constraint were in the fabricated components company, at least some of the problems regarding obsolescence, scrap and volatility would disappear (Figure 14.13).

Notice that the constraint is now at resource 7 in the fabricated components company. This means that the electronics components factory has additional capacity to provide products that may be scrapped or delayed prior to arriving at the computer company. In fact, if the constraint is in the fabricated components company, the electronics components will be waiting at resource 2 before the arrival of the matching fabricated parts.

USING THE EXCESS CAPACITY

The implementation of the DBR process at the vendor generates excess capacity. Used correctly, this excess capacity can be a strategic weapon to protect and enhance the strength of the entire supply chain. As inventories decline and the weak portions of the system become properly buffered, excess capacity can be sold into segmented markets, thereby generating additional revenue. This additional revenue can be used to increase profits or used as a trade-off to meet the customer's demand for lower prices.

THE ADVENT OF THE MULTI-ENTERPRISE INFORMATION SYSTEM

Supply chain management offers some interesting opportunities for software vendors. It would be interesting to see a software project that spans many diverse companies all implementing schedules from the same TOC based information system. The products of many different companies would all have to be placed within a single "net" (see Glossary). Such a project could even be the catalyst for creating service companies whose sole purpose is to schedule the supply chain, as opposed to one company within the chain having that responsibility (Figure 14.14).

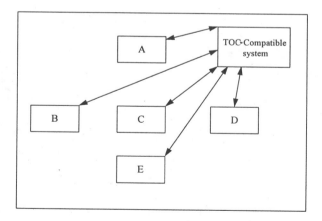

Figure 14.14 The TOC-based information system in supply chain management.

CONCLUSION

Cooperative supply chain management among multiple companies is an important issue and is one of the many "new frontiers" for TOC that will require additional study and implementation. It can help many companies take the next step toward meeting the competitive challenges of today. Key issues for supply chain management include:

- Implementing the decision process within the multi-enterprize environment
- Developing the multi-enterprise information system
- Implementing the five-step process within the multi-enterprize environment

STUDY QUESTIONS

1. What three major influences define how companies react to supply chain management?
2. Explain the supply chain dilemma and how it manifests itself.
3. Why are supply chains of multiple companies the same as any other chain of events?
4. What issues should be considered when placing the constraint?

15

Implementing TOC in a Manufacturing Environment

This chapter will explore the process by which TOC should be implemented.

OBJECTIVES

- To establish the basic sequence of events that will ensure success
- To create an understanding of what activities must be performed and why
- To create an understanding of what to avoid during the process

THE IMPLEMENTATION OVERVIEW

It is important to grasp the scope of the TOC implementation program. The following is a list of issues that should be addressed. Although the order shows a basic flow, the exact sequence is flexible.

- Establish the organization structure and rewrite the corporate policy book. (Note that rewriting the policy book to the TOC world will require a paradigmatic shift in thinking, best accomplished through the education process.)
- Develop the action plan for implementation of the five steps through utilization of the TOC thinking process identifying what to change, what to change to, and how to cause change.
- Establish TOC financial systems and begin collecting data and making decisions based on Throughput, Inventory and Operating Expense.
- Analyze the dependent-variable environment by developing product flow diagrams.
- Establish the QFD process by determining the definition for the necessary condition of good quality and extending it to the production process.

- Identify critical resources and develop the shop floor layout.
- Establish the DBR process and begin buffer management.
- Identify and implement appropriate feedback mechanisms where necessary.
- Establish the supplier certification program.
- Document the process by creating the quality assurance, functional procedures and specifications manuals.
- Begin focusing on productivity. Plan location of constraints and methods of subordination.
- Begin the strategic planning process to recession-proof the company.
- Establish a permanent education program to support the requirements established under QFD and to maintain the TOC improvement process.

BUILDING THE ORGANIZATIONAL STRUCTURE AND REWRITING THE POLICY BOOK

The first step in building the organizational structure and rewriting the policy book is to begin the process of education. There must be a shift in thinking for those who are involved. Unless this shift takes place, the new organizational structure will be in name only. Employees, regardless of job level, will not do what they do not believe in. They must begin the process of leaving the cost world behind completely by rewriting their informal policy book. Once this is accomplished, the organization can be built, the formal policy book rewritten and the development of the action plan begun.

The executive and quality management councils determine what the organizational structure should be, guided by whether or not a program of continuous product improvement is to be endorsed and by what amount of support will be required to implement the action plan. While the functional organizations of marketing, production, purchasing, and so on will remain the same their focusing mechanisms and responsibilities will have to be modified.

BUILDING THE ACTION PLAN FOR IMPROVING PROFITS

Each company is different with regard to its physical environment and policies, requiring different activities in order to improve. While there are certain aspects of each process to be implemented that are similar, the overall scope, tasks, sequencing of tasks, responsibilities, and the distribution of information will be different for each company. To develop an action plan that is effective for a specific company, the prerequisites for success should be known. From this list of prerequisites, an action plan is developed. The action plan represents the implementation plan.

The prerequisite tree is created after a determination has been made to make a specific change in policy or after the method of exploitation and subordination of the constraint has been selected. As a prerequisite to creating the action plan, the current reality tree, evaporating cloud, future reality tree, prerequisite tree and transition tree functions need to be addressed. The implementation plan follows analysis of the physical and policy constraints.

Physical Constraint

To identify the physical constraint(s), it is necessary to

1. Identify the limiting resource
2. Determine how to get the maximum amount of Throughput from it
3. Determine what all other resources must do to protect the maximum amount of Throughput" being generated
4. Determine what skill sets are required and who must be motivated
5. Develop the action plan

Policy Constraint

To identify the policy constraint(s), it is necessary to

1. Identify what policy constraint needs to be changed (current reality trees)
2. Identify what new policy to adopt (evaporating cloud/assumption model)
3. Determine the impact of the change (future reality trees)
4. Determine and solve intermediate obstacles to achieving the goal (prerequisite tree)
5. List the actions necessary to accomplish the prerequisites (transition tree)

Physical constraints can be identified through the use of either current reality trees or TOC-compatible software. As seen in Chapter 13, TOC-compatible software is capable of generating tremendous insight as to the relationship of all resources and the impact of a change before it is made.

ESTABLISHING THE PERMANENT EDUCATION PROGRAM

Obviously, statistical process control and design of experiments are necessary skills needed by shop people and engineering to gain and maintain control over specific processes. However, to accomplish the kind of paradigmatic shift advocated in this book will require a different type of motivation. There are two issues that must be overcome: fear and the "not invented here" syndrome. Change causes fear, which gives rise to resistance. This resistance can paralyze any successful attempt at implementation, specifically if excess capacity is seen as a problem and not a weapon. The "not invented here" syndrome can be just as disastrous. Whenever one portion of the organization has been successful at breaking the constraint so that it moves to another function, there will be a certain amount of resistance from those people in the function at which the new constraint now resides. This situation is made worse when those people who have just broken the constraint try to get the people in the new constraint area to implement the five-step process. To reduce the impact of these two problems, personnel should be encouraged to invent their own solutions. People who have been allowed to do this will have very little fear of change or "not invented here" resistance. This kind of "realization training" can be accomplished by simulation in which the student is placed in an environment where a decision must be made and allowed to see the impact of that decision. Simulations are more like "real life" situations than are textbook examples. However, a simulation cannot be created for all situations, nor is it always desirable to rely on a simulation to motivate people. The motivator must instill a thorough understanding of the cause-and-effect relationships concerning the problem as well as the assumptions involved. Being a group facilitator and leading people through the TOC thinking process helps people to create, for themselves, simple solutions for that they themselves are responsible. Obviously, the facilitator must know something about the process and must have worked through the problem prior to leading the group. The best group facilitators are managers responsible for the groups that lead. Managers should be trained to perform the function of the group facilitator.

Group facilitation is not always desirable. In such a case, the manager must know just the right question to ask to unlock the intuition of the person he or she is trying to motivate. For example, in Chapter 5, quality cost was approached by presenting a specific situation and then asking what the impact would be under obvious conditions, thus making quality cost a questionable focusing mechanism in the mind of the reader. This approach required no real education process. It did require that a thorough study of the TOC thinking process be used regarding quality cost issues.

A balanced education program will incorporate a number of different methodologies to help motivate and to transfer technical know-how including books, workbooks, classroom instruction, simulations and facilitation. However, unlike past education programs that concentrate on a shotgun approach to education, a focused approach should be taken to understand who needs to know what and when they need to know it. The education implementation plan should be incorporated in the overall implementation plan that is defined by the list of prerequisites assigned to the goal. As an example, the constraint has been identified as a lathing operation in the production facility. What are the prerequisites to exploitation and subordination? What obstacles must be overcome? What must occur to maximize the amount of money currently being made from the constraint and what must happen to subordinate the rest of the resources to the way in that it has been decided to exploit the lathe? To maximize what Throughput can be created by the constraint, no defective parts should be made by the constraint, a valid schedule must be created and followed, defective parts should not reach the constraint, engineering should minimize the time used at the constraint to make parts and the product mix should be set to maximize Throughput. What actions should be required to see that each person is able to fulfill the requirements? Persons must not be blocked by thinking about exploitation in an illogical manner. Each person must also understand how to exploit. Is the quality engineer capable of establishing a capability index for the constraint and implementing an effective statistical process control monitoring system? Can production be convinced not to deviate from the schedule? Can engineering implement a properly focused setup reduction program? Only when these questions are answered can an implementation plan can be created.

If education is to be used to motivate people and to enable them to invent their own solutions on a day-to-day basis, it must be a permanent part of the overall program and be available on a regular basis. In addition, the ability to simulate the actual environment will greatly increase the amount of knowledge that can be transferred. The investigative work that can be accomplished through diagnostic simulations yields unique benefits and should be considered part of the overall learning process.

ANALYSIS OF THE CURRENT ENVIRONMENT

Understanding the way that resources interface so that the proper actions can take place is the subject of the product flow analysis and should be one of the first activities performed in an attempt to understand the company. Product flow diagrams are created that document each step in the way products flow through the facility (see Chapter 3).

THE SURVEY

It is important to have knowledge of where each function within the organization is with regard to acceptance and understanding of the tasks to be performed and what the requirements are from two perspectives: understanding what must be changed and determining how the change will occur. Most change can occur after a person (1) understands that a need to change exists and (2) feels able to make the change occur. The survey is designed so that a manager can understand what skill sets and procedures are lacking. A corrective plan of action can then be made. However, there are other perspectives that must be considered. A survey tends to look at the magnitude of a problem without consideration for actual problems connected to the five-step process and tends to cloud the issue by diluting the overall effort. A survey should be conducted only after the constraint has been identified and a determination has been made of how to exploit and subordinate. It is important to know whether the proper skill sets are available to successfully accomplish the exploitation and subordination phase?

ESTABLISHING EMPLOYEE INVOLVEMENT TEAMS AND QUALITY CIRCLES

The first step in the development of an employee-involved TOC/TQM program is to establish the concept of a valid focusing mechanism through education. The next step is to reinforce the first by changing the employees' environment, the methods by that feedback is received and the way individual work goals are established:

- Establish the employee involvement and QC team concept through education in TOC/TQM concepts and elementary tools, including statistical process control, Pareto analysis, current reality trees, assumption modeling, future reality trees, prerequisite trees and transition trees, as well as DBR concepts, buffer management and brain storming
- Clear and simplify the work environment by designating a place for everything, and keep the work area cleared of any tool or product that is not currently being used
- Establish rules for activating resources and determining priorities, and ensure that priorities are easily determined
- Establish feedback mechanisms including source inspections (next process and poka yoke) and the buffer management process, where necessary, to support the exploitation and subordination processes
- Begin regular meetings to improve communication, reinforce TOC concepts and determine what actions must be taken to exploit the constraint and to subordinate the remaining resources
- Document the process

IMPLEMENTING DRUM-BUFFER-ROPE

The DBR implementation process begins with education. Specific educational requirements include understanding

- How resources interface
- How a schedule should be generated
- What the rules of resource activation are
- How material releases are determined
- What the five-step process of continuous profit improvement is
- How to exploit the constraint
- How to properly subordinate
- How to make decisions

The next step is to analyze and establish relationships among resources to build an understanding of the overriding characteristics of the manufacturing facility and determine whether the problems encountered are characteristic of an A-plant, T-plant, V-plant or a combination of each.

Once the plant type has been determined and the constraints have been located, rules for resource activation and determining priorities can be set. Shop documentation will be required for maintaining order identity and continuity. While the traditional work order is no longer needed for dispatching and determining priorities or to support cost accounting, the traditional MRP II work order system generates shop paper for material kitting and routing, and supports the reporting of material movement through the shop.

Additional documentation includes

- A schedule for all constraint resources
- A schedule for resources that appear immediately after a diverging operation to prevent stealing of material
- A raw materials release schedule to coordinate the release of material with the constraint schedule
- A raw material requirements report to notify purchasing of the quantity and need date
- A sales order due date report to keep track of customer commitments.
- A buffer management report to execute the buffer management process.

While documentation is being generated, the shop floor layout can be rearranged to ensure ample spacing for buffer inventories and reduce the amount of space for nonconstraint resource inventories.

Once the shop paper has been completed, a schedule for the constraint should be generated to set priorities for production and to determine the requirements of all other resources through the development of the rope (see Chapter 6). An estimate of the amount of aggregated protection necessary in front of the various buffer origins including the shipping, constraint and assembly operations will be needed to determine the release dates for raw material. Buffer sizes should be oversized estimates until some indication of the performance of buffer zones is available. As more information is obtained about the makeup of each zone cross section, the size of the buffer can be increased or decreased as necessary (see Chapter 6).

Most manufacturing facilities start with excess material in work in process (WIP). This material can be downsized through attrition, or it can be completely withdrawn from the production floor. Obviously, obsolete material should be totally withdrawn. Production batch sizes should be determined by the sales order or finished goods forecast. The planning of overlapping operations can be accomplished by first determining what the size of the transfer batch should be and then by moving material by the transfer batch size. The smaller the transfer batch, the shorter the lead time will be.

It may be convenient—but not necessarily required—to colocate resources into cells that use the same parts/operations repetitively so that transfer batch sizes can be held to a minimum. Remember, however, that improvements should be made so that profit goes up. Cells do not guarantee that profits will go up, nor does a transfer batch size of 1.

A cell is a close assemblage of operations that cuts down on distance traveled and the support necessary to move material. In the U-line, one form of cell, the operator stands in the middle of the "U" and may operate more than one machine.

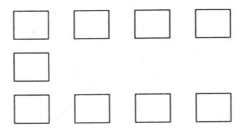

At this time, the constraint schedule can be generated and the material release schedule created along with the buffer management reports so that production can begin. It is important to document the production process and the rules of resource activation and improvement so that workers can have a ready reference.

CREATING THE TOC FINANCIAL SYSTEM

The TOC financial system is simple and very straightforward. The past, current, and predicted Throughput, Inventory and Operating Expense figures should be understood. Total Throughput is equal to total sales minus the total cost of raw material. Since incremental cost buildups are unable to reflect the true value of any product, the labor and overhead figures for assemblies, sub-assemblies and finished products can be eliminated. The cost of goods sold figure for most accounting systems would then reflect only raw materials. Most systems also reflect sales price. The profit margin figure, which is equal to the sales price minus cost of goods sold, represents Throughput regardless of whether one is looking at past history, current status or forecast. This approach may cause a problem for taxation authorities or corporate offices, which have very definite opinions as to how inventories should be valued. It may be necessary to keep two sets of books, one for auditing and one for making decisions and understanding the true health of the company.

Inventory includes the value of raw material contained in raw material, work in process and finished goods inventories, but it also includes the assets of the corporation that cannot be easily withdrawn such as machines, buildings, desks and computers. Operating Expense includes the day-to-day expenses for any type of labor, insurance, office supplies, taxes and utilities. A separate report can be generated along with separate major accounts for all three account types in most accounting software packages, and current accounts can be grouped under major accounts for Throughput, Inventory and Operating Expense.

GETTING STARTED

In the process of trying to determine what to fix and how to develop the implementation plan, it is sometimes difficult to know where to start. It is important not to lose sight of the objective, which is to increase profitability by implementing the five-step improvement process. The first step is to identify the constraint. In accomplishing this first step, it is often useful to try to isolate the constraint by function. Sample questions to propose might be:

If you had to deliver 20% more product from manufacturing could you do it?

If not,

Would you have to obtain more resources from outside the company?

This will determine whether the constraint is inside the company or outside in the market. If the constraint is in the market and the answer is "Yes, we could deliver," then ask

Why aren't you selling more?

and begin to assemble the current reality tree to look for the core cause. Or else, one would ask

How can we get more out of the market?

and begin the exploitation process. If the constraint is internal to the company and the answer is "No, we cannot deliver," one would ask

Why can't we deliver more?

Another approach is to begin assembling 6–10 undesirable effects and to begin the TOC thinking process by generating current reality trees, evaporating clouds, future reality trees, prerequisite trees and transition trees.

If a physical resource is found to be acting as if it were the constraint, one can try to determine how to exploit it by building a schedule for it to maximize its availability. It may be found during this process that the constraint is not physical but, due to the way in which the suspect resource is being managed, is nevertheless limiting the creation of Throughput.

After the constraint has been exploited, one should determine what needs to be done by all the other resources to support the constraint. At this point, it may be found that other resources are also restricted, and because of a restriction elsewhere it may be difficult to exploit the constraint to its fullest. One may need to discover and exploit the secondary constraint. In squeezing the maximum out of the secondary constraint, it is important

- To subordinate the secondary constraint to the primary constraint
- To manage the secondary constraint when the primary one is broken

For example, the market may be the constraint. But the manufacturing manager states that because of a limit in production he cannot deliver 20% more product, even if it could be sold. However, he says he could deliver an additional 5%. In this case, the market would be considered the primary constraint, while production would be the secondary constraint. Because the constraint is in the market, it is extremely important that all orders be delivered on time. Production must be subordinated to the market. That resource that is considered the secondary constraint should be exploited to ensure that it can be subordinated to the market. A schedule should be created for the limiting resource, and an improvement program should be implemented to maximize its ability to deliver to the schedule created by the market. All other applicable resources should be subordinated to the secondary constraint. These steps will help guarantee that the secondary constraint and all other resources are prepared for the day when the primary constraint is elevated.

While attempting to exploit the market, it may be found that a basic factor such as cost is preventing the maximum amount of Throughput from being generated. Once market segmentation has been accomplished and excess capacity has been sold, the constraint will immediately move inside the company. More than likely, it will move to the secondary constraint. Once this occurs, the marketing strategy must change to ensure that the internal constraint is fully exploited. One can then begin to understand what must be done to properly subordinate all resources to the new constraint.

It is very important to foster creative approaches. Addressing the improvement process by using the five steps will help tremendously in being able to focus the thought process and will promote creativity.

STUDY QUESTIONS

1. What issues should be addressed in the implementation of TOC?
2. What is the permanent education program and why is it needed?
3. What are the education requirements for the DBR process?
4. What is the TOC financial system and how can it be created from the traditional system?

16

Facing the Strategic Issues

This chapter addresses some key issues with respect to strategic planning and offers insights into how to help a company grow and protect its ability to create profits.

OBJECTIVES

- To create an understanding of market segmentation strategy and its overall implementation
- To create an understanding of how to make a company grow and the importance of the utilization of resources in the creation of Throughput as well as in the support of the market segmentation strategy
- To create an understanding of the concept of long-term planning as it applies to the five-step improvement process and the impact of introducing new products
- To begin to understand how to recession-proof a company
- To understand the impact of the farmout strategy on profitability
- To define the role of the diagnostic system

MARKET SEGMENTATION

Possibly the most important and yet least understood tool in strategic planning is effective market segmentation. Market segments are designed to maximize the use of resources in the production of Throughput and to minimize the risk of a downturn. There are three rules in market segmentation. Market segments should be selected so that

1. The sales price in one market will not impact the sales price in another
2. The same resource base serves all markets
3. It is unlikely that all segments will be down at the same time

The benefit of an effective program is the ability to protect valuable resources and to maximize profitability by manipulating the market to fit capacity instead of manipulating capacity to fit market demand. By manipulating the sales price for products that are made at nonconstraint resources, it is possible to increase product demand and thereby increase demand for time at those resources that are being underutilized. Since the real profit for these products is the difference between the sales price and the cost of raw material, a large amount of profit can still be generated during those times when there is a downturn of demand at some resources.

There are many ways to segment the market: geographical, product, time, customer type—the list is very large. Probably one of the most effective is product segmentation where two products are similar but, with minor changes, can serve different markets and demand different prices. Even when product configurations are more complex, it may be more profitable to sell them at a lower price to a different market if the additional products are produced at resources that have excess capacity. A good example of this is in the computer industry, where a basic product is sold at one price and the same product with minor add-ons is sold at a lower price to the same customer base.

Geographic market segmentation involves the sale of products to two different geographical locations. The product can be the same exact part but because of a difference in demand is sold at a lower price at one location to fill excess capacity.

A market segmentation based on time may segment prices to lead times and take advantage of a lead time that is shorter than the industry average. The average lead time for a company to produce a circuit board for its customers may be one week, while the industry average may be six weeks. The company may set its prices based on the industry average lead time of six weeks. If the customer wants the product in two weeks, it will have to pay a premium. While it does not cost the board manufacturer any more money to produce the board on the 2-week schedule, the company can manipulate the price for earlier delivery to increase throughput and to protect its resources.

Product pricing is not the only factor that can be manipulated through market segmentation. In segmented markets where excess capacity is being sold, what would be the impact on product acceptance if the additional profit being generated were used to increase the quality or functionality of the products being produced? The following figure illustrates this.

Product	M	L	O	STD
AFG	80	60	120	260

$300	$200	$100

In the $200 market, while selling excess capacity, the profit will be the difference between the $200 sales price and the $80 raw materials price. What would be the impact on product acceptance in the $200 market if $20 of additional raw material were added, making the total raw materials investment $100? Or if 5-10 hours of additional labor were used to enhance the product?   If this is achieved with excess capacity, the profit is still $200 minus the $100 cost of raw material. This strategy would be a way of increasing product quality, decreasing the sales price and gaining market share and would provide a tremendous competitive advantage.

MAKING THE COMPANY GROW

A company can grow by implementing the five-step improvement process, but at what point does it know when to add employees, buy new resources or build new plants? New plants are added when there is no capacity left in the old plant. Only when everything has been done to squeeze as much Throughput as possible from the old plant and all resources are being fully utilized, not just activated, should consideration be given to buying a new plant.   The question must be asked whether the resources in the new plant can be protected through adequate market segmentation. If the answer is no, then an effort must be made to look for new segments to support the building of the plant. When this question can be answered satisfactorily, the plant can be built.

Employees should be added with the idea that they are a permanent fixture. Whenever employees are added, it is the responsibility of management to do everything within its power to keep them.   Employees are added so that Throughput can be protected or additional Throughput generated as part of the five-step process. They are kept by ensuring that resources are fully utilized in the creation of Throughput.   If employees are led to believe that their efforts to increase the amount of excess capacity available through the improvement process will be used not to protect but to threaten their jobs, the improvement process will become threatened. In addition, whenever employees are laid off, they may get jobs in companies where their expertise and experience levels are most in line with the job requirements. Those jobs will be at competitive companies. When a major personal computer manufacturer that had broken all records in growth during its first few years in business was having financial difficulty for the first time, it fired its president and laid off an estimated 1,000-1,400 people. Within a week a major competitor had ads in the local newspaper seeking to hire them. What better way to pick up good people trained by one of the best computer manufacturers in the world.

Much like employees, machines are added when necessary in the implementation of the five-step process. If a machine becomes the constraint and it is elevated, Throughput should go up. However, there are a number of conditions for which the best option is to maintain the constraint at its current location. Whenever, a constraint moves to a machine that is processing parts erratically because of poor quality or increased downtime, the production process becomes unpredictable. Constraint resources that feed themselves in the process will cause Inventory to begin climbing and increase the amount of unpredictability. Before a machine is purchased to elevate the constraint, the location of the new constraint should be known and its impact predicted.

It may also be necessary to buy a machine to subordinate the process. If a resource is a secondary constraint and is loaded to 90% capacity, it may be impossible to subordinate it to the primary constraint. In this case, a resource is elevated to smooth out the scheduling process and maintain predictability.

Companies can grow without adding resources by building up the strong links of the chain. This is the objective of a market segmentation program.

LONG-TERM PLANNING

Whether for the long or short term, the objective of planning is to live and profit by the limitations provided. Long-term planning can involve breaking the current constraint; elevating specific market constraints, which may take a long time; or manipulating and predicting where the constraint will appear next. If the constraint is a specific machine within production and it is supporting a certain level of profitability, what will be the impact of breaking the constraint? Where will it appear next and what actions will be required? If it goes into the market, how should the five-step improvement process be implemented? To simply wait until the last constraint is elevated and then find out where it has gone is like playing Russian roulette. In some instances, the identification process should simply be a verification that the constraint is where it was predicted to be.

NEW PRODUCT INTRODUCTION

Whenever new products are introduced into a company, there is a certain amount of disruption for those resources that are controlling the generation of Throughput. New products that take time from the constraint also diminish Throughput. Managers should be aware that unless the Throughput generated per unit of the constraint is equal to or higher than current production, Throughput will go down. The impact of new products can be minimized during the design process by planning the use of nonconstrained resources for production. This will allow the company to maintain a certain price advantage during the introduction period and also to generate higher levels of Throughput during a critical period when the expenses for new product development and market introduction are high. Disruptions in the schedule will be nonexistent since nonconstraint resources will still be able to produce to the schedule presented by the constraint. Since this is also a period of uncertainty for the new product, risk is held to a minimum. even a certain amount of the cost of production, namely the price of raw material, can be recovered. In situations where it is impossible not to use a specific resource that is the constraint, new products should minimize its use. An alternative would be to move the constraint to a new location where the Throughput generated would be higher and the restrictions to the new product would be less serious.

In many cases, the introduction of new products will increase the load on non-constraint resources to the extent that they will be less able to deliver to the constraint on time, requiring that buffer sizes be increased or that overtime be used to maintain current schedules. While Throughput may not suffer, Inventory and Operating Expenses will go up. This should be considered in the expense of introducing a new product.

RECESSION-PROOFING THE COMPANY

Companies move into stagnation and decline when they fail to implement the five-step improvement process. In a recession the continuation of this process is more difficult due to the shrinking market size. Dealing with issues of marketing constraints is not something most companies are comfortable with. However, the problem is not insurmountable.

Recession-proofing enables a company to continue growing regardless of the economic status of the environment in which it exists. The company must demonstrate:

- The ability to continuously implement the five-step improvement process
- The ability to minimize the impact of a recession by protecting resources
- A valid decision process

It is imperative that companies understand how to consistently identify, exploit, subordinate, elevate and repeat the process, regardless of the situation. The better a management group is at this process, the more isolated from threat the company becomes.

Whenever companies enter a period of decline, there are two events that seem almost inevitable: the layoff and the reorganization. The first occurs to satisfy the immediate concerns of the stockholders, namely to increase profitability by reducing cost. The second occurs to either consolidate operations so that costs can be further reduced or to segment resources in an attempt to increase control, thereby increasing future Throughput. However, Operating Expense exists to support and protect the creation of Throughput. Before any measure is taken to reduce Operating Expense a determination as to the identity of the constraint as well as the method of exploitation and subordination must be made. Reducing Operating Expense before knowing what will be required to elevate the constraint is dangerous. In the section on decision making, product pricing was discussed. The importance of product pricing and using excess capacity in gaining a competitive advantage of pricing in a specific market should be clear. If the method of exploiting the market is to find an isolated segment and dump excess capacity at a reduced price, reducing excess capacity will reduce the effectiveness of this alternative by reducing the amount of money that can be generated.

Segmenting resources in the face of a marketing constraint usually means that the same market must absorb more Operating Expense. In recession-proofing a company, the objective is to have a large number of market segments being fed by the same resources, thereby protecting those resources that are available. Resource segmentation or plant expansion should occur only when no excess capacity exists in the current plant and when enough market segments exist so that even if one market fails, others are there to absorb available resource capacity. If a marketing constraint and excess capacity exist at the same time, the rule is to find a market segment and dump products onto that market for any amount one can get above the cost of raw material.

Without a valid decision process based on absolute measurements and an understanding of the relationships among resources, all the preparation in the world will not help to avoid the inevitable. Companies must try to react to the environment instead of molding it to fit their needs.

THE FARMOUT STRATEGY

There are many ways to commit corporate suicide, but one of the most devastating and probably the most prevalent is the farmout strategy. Companies who are having a tough time competing and need a quick solution invariably begin comparison shopping between the cost of doing business in-house and the cost of buying parts and subassemblies from vendors and assembling them in-house. Unfortunately, the short-term winners tend to be the vendors who supply the parts, not the corporation who gave them the business. The corporation that farmed out the material begins to lose immediately, and in the long run, everyone loses. As the company begins to farm out products, the labor and overhead assigned to the parts now coming from the vendor must be distributed to the parts that are still being built in-house. The overhead distribution becomes larger and larger until it appears that no parts made in house can compete with vendor prices. The following figure is a cost buildup for two similar parts made in house.

	Material Cost	Labor Cost	Overhead Cost	Standard Cost	Sales Price	Profit Margin
A	80	60	120	260	300	40
B	80	80	160	320	310	−10
					Net Profit	30

Under the cost matrix, it looks as if part B is not profitable and a solution must be implemented that will bring down costs for B. A vendor that is ready to supply B for $200 is located. Initially, the solution seems to work. However, unless the total overhead and labor force is reduced, Operating Expense will not go down. The labor and overhead must now be placed solely on A.

	Material Cost	Labor Cost	Overhead Cost	Standard Cost	Sales Price	Profit Margin
A	80	140	280	500	300	−200
B	200				310	+110
					Net Profit	− 90

This may seem a trivial issue, but it has resulted in the downfall of many corporations. The make/buy decision is considered in Chapter 5, on correcting the decision process.

The decision as to whether a product should be farmed out should be made based on its impact to Throughput, Inventory an Operating Expense, not on its cost. And the primary focus should remain on Throughput.

CONCLUSION

Strategic planning can have a tremendous impact on any company. Understanding the concept of market segmentation, recession-proofing, long-term planning, new product introduction and how to make a company grow is extremely important. Mastering these techniques will help companies to grow fast and stay healthy.

STUDY QUESTIONS

1. What are the three rules of market segmentation?
2. What advantages does market segmentation provide?
3. List and define 5 methods of market segmentation?
4. What disadvantage can be created by eliminating excess capacity as it is generated through the improvement process?
5. When is it appropriate to add new resources?
6. What key issues should be considered during new product introduction?
7. How should excess capacity be exploited?
8. What are the fallacies of the farmout strategy?

Appendix
Normal Table and F Table

Table A.1   Normal Table

Z		.00	.01	.02	.03	.04	.05	.06	.07	.08	.09
3.2	3	68714	66367	64095	61895	59765	57703	55706	53774	51904	50094
3.3		48342	46648	45009	43423	41889	40406	38971	37584	36243	34946
3.4		33693	32481	31311	30179	29086	28029	27009	26023	25071	24151
3.5		23263	22405	21577	20778	20006	19262	18543	17849	17180	16534
3.6		15911	15310	14730	14171	13632	13112	12611	12128	11662	11213
3.7		10780	10363	99611	95740	92010	88417	84957	81624	78414	75324
3.8	4	72348	69483	66726	64072	61517	59059	56694	54418	52228	50122
3.9		48096	46148	44274	42473	40741	39076	37475	35936	34458	33037
4.0		31671	30359	29099	27888	26726	25609	24536	23507	22518	21569
4.1		20657	19783	18944	18138	17365	16624	15912	15230	14575	13948
4.2		13346	12769	12215	11685	11176	10689	10221	97738	93447	89337
4.3	5	85399	81627	78015	74555	71241	68069	65031	62123	59340	56675
4.4		54125	51685	49350	47117	44979	42935	40980	39110	37322	35612
4.5		33977	32414	30920	29492	28127	26823	25577	24386	23249	22162
4.6		21125	20133	19187	18283	17420	16597	15810	15060	14344	13660
4.7		13008	12386	11792	11226	10686	10171	96796	92113	87648	83391
4.8	6	79333	75465	71779	68267	64920	61731	58693	55799	53043	50418

Table A.2   F Table

DF	1	2	3	4	5	6	7	8	9	10
1	161.44	199.50	215.69	224.57	230.16	233.98	236.78	238.89	240.55	241.89
2	18.51	19.00	19.16	19.25	19.30	19.33	19.35	19.37	19.39	19.40
3	10.13	9.55	9.28	9.12	9.01	8.94	8.89	8.85	8.81	8.79
4	7.71	6.94	6.59	6.39	6.26	6.16	6.09	6.04	6.00	5.96
5	6.61	5.79	5.41	5.19	5.05	4.95	4.88	4.82	4.77	4.74
6	5.99	5.14	4.76	4.53	4.39	4.28	4.21	4.15	4.10	4.06
7	5.59	4.74	4.35	4.12	3.97	3.87	3.79	3.73	3.68	3.64
8	5.32	4.46	4.07	3.84	3.69	3.58	3.50	3.44	3.39	3.35
9	5.12	4.26	3.86	3.63	3.48	3.37	3.29	3.23	3.18	3.14
10	4.96	4.10	3.71	3.48	3.33	3.22	3.14	3.07	3.02	2.98
11	4.84	3.98	3.59	3.36	3.20	3.09	3.01	2.95	2.90	2.85
12	4.75	3.89	3.49	3.26	3.11	3.00	2.91	2.85	2.80	2.75
13	4.67	3.81	3.41	3.18	3.03	2.92	2.83	2.77	2.71	2.67
14	4.60	3.74	3.34	3.11	2.96	2.85	2.76	2.70	2.65	2.60
15	4.54	3.68	3.29	3.06	2.90	2.79	2.71	2.64	2.59	2.54
16	4.49	6.63	3.24	3.01	2.85	2.74	2.66	2.59	2.54	2.49
17	4.45	3.59	3.20	2.96	2.81	2.70	2.61	2.55	2.49	2.45
27	4.21	3.35	2.96	2.73	2.57	2.46	2.37	2.31	2.25	2.20

Glossary

Action. An activity defined in a transition tree which is used to obtain an intermediate objective.

Activation. The employment of a nonconstraint resource for the sake of keeping busy. Activation does not imply usefulness in supporting system Throughput (APICS).

Additional cause. In the TOC thinking process, a secondary cause that is identified when one cause by itself is insufficient to explain the existence of an effect.

A-Plant. A plant characterized by a large number of converging operations starting with a wide variety of raw materials items being assembled in succeeding levels to create a smaller number of end items.

Aggregation of demand. The result of the use of the formula Capacity divided by Demand, in which all demand is aggregated into time buckets. The assumption is made that all capacity is available to all demand within the time bucket.

Artificial demand. A feature of the dynamic buffering process originating from internal processing logic. The demand placed at a resource to protect available capacity when a certain number of consecutive periods on a resource have been scheduled at 100% by external demand.

Ascending parts/operations list. A feature of the TOC-compatible information system that lists individual parts or operations in ascending order from raw materials to the final product.

Assembly buffer. Used to ensure that those assembly operations directly fed by the constraint do not wait for materials to arrive from nonconstrained legs within the net. It also determines the release schedule for raw material into those operations that feed the assembly buffer.

Assembly schedule. Used to sequence the arrival of parts to an assembly operation from nonconstraint operations. Assembly schedules are used only when at least one leg of the assembly operation is being fed by the constraint(s).

Backward rods. Rods that require an order to be scheduled across a constraining resource a given amount of time after a previous order at the same or a different resource.

Batch rods. A time mechanism used to separate/protect orders (batches) on the same resource schedule.

Behavioral constraint. A behavior that is in conflict with reality and blocks the exploitation of a the physical constraint.

Bottleneck. Anytime the demand placed on a resource is equal to or more than capacity.

Budget. A performance measurement used to relate the forecasted expenditure requirements at the departmental level. This information is used to predict and control Operating Expense and should not be used to focus improvement.

Buffer. A time mechanism used to protect those portions of the factory (buffer origins) that are vulnerable to problems associated with statistical fluctuations.

Buffer management. A technique used to manage the amount of protection necessary and the process of controlling the buffer origins within the plant

Buffer management report. A tool for consolidating the input of the buffer management worksheet to determine the relative impact by injecting the Throughput dollar days and Inventory dollar day equation.

Buffer management worksheet. A tool for organizing and facilitating quantitative analysis and to ensure uniformity in data collection. The buffer management work sheet is used by the buffer manager to collect and assemble data relevant to those orders that require expediting.

Buffer manager. The person designated to manage the buffer and ensure that the buffer origin is protected by expediting material causing holes in zone 1 of the buffer.

Buffer origin. That portion of the system needing protection and the object of the buffer. Buffer origins include but are not limited to the constraint, secondary/ tertiary constraint(s), shipping and those assembly operations having one leg being fed by the constraint.

Buffer system. The total system used to protect the constraint from the impact of statistical fluctuations

Buffer zones. The three zones within the buffer management process. See also **Tracking zone** and **Expedite zone**.

Capacity. Usually expressed in hours minutes or quantities for a given period of time, the capability of a resource to react to demand.

Capacity-constrained resource. A resource that has, in aggregate, less capacity than the market demands.

Categories of legitimate reservations. Used in the TOC thinking process, these question the existence of a specific entity, its cause, or the relationship between them.

Cause insufficiency. Used in the TOC thinking process to identify a situation in which the suspected cause is insufficient to explain the result.

CCR. See **Capacity-constrained resource**.

Constraint buffer. Used to protect the constraint(s) and to determine the release of raw materials to those operations that feed the constraint.

Constraint management. The practice of managing resources and organizations in accordance with principles of the Theory of Constraints (APICS).

Combination plants. Manufacturing plants that have more than one of the VAT characteristics. See also **V-Plant**, **A-Plant** and **T-Plant**.

Combination rods. Rods that require an order to be scheduled across a constraining resource a given amount of time before and after a previous or succeeding order at the same or a different resource.

Constraint. Anything that limits the system from attaining its goal. Constraints are categorized in the following manner: behavioral, managerial, capacity, market, and logistical.

Constraint schedule. The schedule created for a capacity-constrained resource in order to exploit its productive capability.

Continuous profit improvement. The five-step process designed to continuously improve the profits of the company. The steps include identifying the constraint, exploitation, subordination, elevation, and repeating the process.

Cost mentality. The tendency to optimize local measurements at the expense of global measurements.

Converging operations. Operations that receive parts from more than one part/operation.

Currently reality tree (CRT). Used to find the core cause or causes from undesirable effects (UDEs).

Dependent setup. Used to maximize the utilization of the constrained resource by combining setups on orders for two different parts/operations having the same setup requirement.

Dependent-variable environment. The environment is the environment where resources are dependent on each other in their capability to produce and are subject to variation in that capability.

Descending parts/operations list. A feature of the system which lists individual parts/operations in descending order.

Desirable effect. Used in the TOC thinking process in the creation of the future reality tree to identify those effects that are beneficial and that are created as a result of an injection.

Diverging operations.   Operations that deliver parts to more than one part/operation.

Drum.   The schedule for the primary constraint that establishes the rate at which the system generates Throughput.

Drum buffer rope.   A scheduling technique developed using the theory of constraints.   The drum is the schedule for the primary constraint and establishes the rate at that the system generates Throughput.   The buffer is a time mechanism used to protect those places within the schedule that are particularly vulnerable to disruptions.   The rope is the mechanism used to synchronize the factory to the rate of the constraint and determines the release date for material at gating operations.

Dynamic buffering.   A method used to improve the buffering process so that overall buffer sizes can be shrunk.   The buffer is allowed to grow when increases in resource demand require additional lead time.   Dynamic buffers shrink as demand declines.

Dynamic data.   Data that is allowed to change as the environment and reality change.

Effect-cause-effect.   An earlier method used in the TOC thinking process that looked for core causes by continually establishing cause-and-effect relationships and supporting each relationship with an additional effect.

Elevating the constraint.   The act of increasing the ability of the constraint in the creation of Throughput by means other than exploitation.

Entity.   Used in the TOC thinking process to identify either a cause or an effect.

Entity existence.   One of the categories of legitimate reservation that questions the existence of an entity.

Evaporating clouds (EC).   Used to model the assumptions that block the creation of a breakthrough solution.

Excess capacity.   Capacity that is not used to either produce or protect the creation of Throughput.

Expedite zone.   Zone 1 of the buffer.   The expedite zone is used to indicate when the absence of an order is threatening the exploitation of the constraint and to notify the buffer manager when to expedite parts.

Exploitation.   The process of increasing the efficiency of the constrained resource without obtaining more of the resource from outside the company.   Exploitation may include, but is not limited to, the implementation of statistical process control, total productive maintenance, setup reduction, creating a schedule, setup savings and the application of overtime on a constraint resource.

First-day peak load.   A load placed at a resource during the subordination process which exceeds available capacity on the first day of the planning horizon.

Forward rods. Rods that require an order to be scheduled across a constraining resource a given amount of time prior to a succeeding order at the same or a different resource.

Future reality tree (FRT). Used to model the changes created after defining breakthrough changes from the evaporating cloud.

Global measurements. Those measurements used to measure effectiveness from outside company (i.e., return on investment).

Glue. A technique used during setup savings to attach orders together for the purpose of scheduling the constraint. Once orders have been attached, they remain attached until the order is complete.

Haystack-compatible system. An information system based on the Theory of Constraints and inspired by the book *The Haystack Syndrome*, by Eliyahu Goldratt.

Holes in the buffer. The absence of orders within zones 1 or 2 of the protective mechanism called the buffer.

Inertia. The tendency to continue looking at a specific problem or activity the same way even though the situation has changed.

Injection. In the TOC thinking process, the change used to eliminate the undesirable effects.

Intermediate objective. In the TOC thinking process, an objective attained to overcome an obstacle defined in the prerequisite tree.

Inventory. One of the key measurements used to manage a TOC company. All the money the system invests in purchasing things the system intends to sell.

Inventory dollar days (I$D). A measurement used to minimize those activities that occur before they are scheduled. It is represented by the formula Inventory quantity × Dollar value × Number of days early.

Leg. When referring to the product flow diagram, or net, a string of parts/operations that feed one part to an assembly operation. An assembly operation has two or more legs.

Leveling the load. The initial attempt at scheduling the constraint. The objective is to place each order on a time line so that there are no conflicts between orders and so that daily capacity is not exceeded.

Local measurements. Performance measures used at the department level to guide activity.

Logical implementation. One of the three phases of implementation of the TOC-compatible information system, involving changes in the way in which the company is managed.

Logistical constraint. A condition that exists when the planning and control system prevents the company from attaining its goal

Managerial constraint. A management policy that prevents or restricts the company from attaining its goal.

Market constraint. A condition that exists when market demand is less than the capability of the company to produce.

Market segmentation. A strategy for recession-proofing and maximizing profit by creating different markets to absorb capacity from the same resource base.

Master production schedule. Used as input in the creation of the net, the combination of the forecast and sales order requirements. It provides the initial schedule from which the identification process used to find the primary constraint is begun.

Moving in time. The procedure used to process orders on the net. Unlike the low-level code used to control the order of processing in material requirements planning, the TOC system uses time. As each order is processed, it is placed in chronological order beginning with the end of the planning horizon. Once all activities have been planned for a specific date, the system moves in time to pick up those requirements of the preceding period. Planning begins at the end of the planning horizon and moves in time toward time zero.

MPS. The master production schedule.

Near-constraint. A resource that is loaded to near-capacity levels. While a near constraint may not limit the amount of Throughput that the system can generate, it can have a negative impact on the subordination process and will result in an increase in Inventory and Operating Expense.

Necessary condition. Boundaries placed on a company's departments or individuals, originating either internally or externally, that serve to regulate activity.

Negative branch. In the TOC thinking process, the reason that a certain activity cannot or should not be done. Negative branches are normally trimmed so that solutions become more robust.

Net. The combination of all product flow diagrams for all products on the master schedule.

99-to-1 Rule. The rule stating that in a dependent variable environment only 1% of the activity is required to produce 99%t of the impact.

Nonconstraint. A resource that has more than enough capacity to meet the market demand.

Nonconstraint schedule. A schedule produced for nonconstraint resources presenting those activities that must not be done prior to a specific time.

Obstacle. In the TOC thinking process, something that will negate or reduce the effect of the injection and its impact on the undesirable effect.

Off-loading. The process of removing specific orders from a constraint resource and placing them at a nonconstraint resource.

Operating Expense. One of the key measurements used to manage a TOC company. All the money the system spends turning Inventory into Throughput.

Overtime schedule.   The schedule for overtime requirements resulting from the exploitation process.   The schedule includes the resource at which the overtime is to be spent, the application date and the amount of time required.

Part/operation.   A combination of the part number from a bill of material and the operation within a routing that is used to produce a specific part.   The smallest element within the product flow diagram.

Peak load.   A period in time where the demand placed on a specific resource exceeds capacity.

Permanent education program.   An education program designed to support the implementation and execution of the theory of constraints.

Physical implementation.   That part of the implementation involving the interface of the user with the system and that includes plant layout, schedule generation, schedule execution and buffer management.

Prerequisite tree (PT).   In the TOC thinking process, a tool used to uncover and solve intermediate obstacles to achieving the goal.

Process batch size.   The number of parts processed at a resource without intervening setup.

Process of continuous profit improvement.   The five-step process used in the Theory of Constraints, including identification, exploitation, subordination, elevation and repeating the process.

Product flow diagram.   A diagram made from the combination of a bill of materials and routing for a single product.   It presents individual parts/ operations and their relative positions within the production process.

Productive capacity.   Capacity required to produce a given set of products, not including that capacity used to protect the schedule from natural fluctuations in capability.

Protective capacity.   Capacity used to protect the capability of a resource or group of resources in meeting the schedule.

Pushing the load.   The process of shifting resource load into later periods.

Pushing the order.   The act of pushing a specific sales order on the schedule into a later time period.

Raw materials schedule.   The schedule produced by the system that defines the quantity and date of raw materials requirements.

Recession-proofing.   A methodology designed to ensure that the company will continue to grow, regardless of the economic environment in that it exists. Included in the recession-proofing process is the ability to: continuously implement the five-step process of improvement, minimize the impact of a recession by protecting resources through proper market segmentation, and make valid decisions.

Red lane.   The string of resources that lead from the primary constraint to the sales order.   The red lane is the most vulnerable portion of the factory.

Release schedule. The schedule for the release of raw material into the gating operations.

Reservation. See **Categories of legitimate reservation**.

Road map. See **TOC thinking process**.

Rods. (See **Time rods** and **Batch rods**.)

Rod violation. A condition that exists when a specific order having forward, backward or combination rods is compromised by a schedule in which another order occupies space too close in time.

Rope. The mechanism used to synchronize the factory to the rate of the constraint and determines the release date for material at gating operations.

Schedule conflict. A situation in that parts/operations or batches scheduled at a given resource are in conflict with other parts/operations or batches due to a lack of adequate protection.

Secondary constraint. A resource whose capacity is limited to the extent that it threatens the subordination of the primary constraint.

Setup savings. The act of combining orders to reduce the number of setups and increase the amount of protective or productive capacity.

Shipping buffer. Used to protect the shipping of finished goods, determine the initial schedule for the constraint and establish the release schedule for raw material that does not go through the constraint or assembly buffer

Shop floor control. That activity associated with controlling the production process and that includes (1) controlling the priority of work (the schedule for the constraints, the release of orders to the gating operations, the schedule for the assembly buffer, the schedule for the shipping buffer and the buffer management system); (2) collecting and feeding order movement, location and quantity information back to the shop floor control system; and (3) measurement tracking (collecting and reporting data for maintaining Throughput dollar days and Inventory dollar days statistics).

Static buffer. That portion of the buffer that is used to protect the buffer origin but that is not allowed to fluctuate.

Static data. Data that is unrelated to the changes that occur in the environment in which it is used.

Statistical fluctuation. The natural variation evidenced in a manufacturing environment by the fluctuation in the ability of a resource to meet a specific schedule.

Station. See **Part/operation**.

Subordination. Providing the constraint what it needs from other resources to maximize the amount of Throughput generated.

Technical implementation. That portion of the implementation dealing with hardware and software, including system selection, integration and testing.

Temporary bottleneck. A situation created when the demand for time at a specific resource temporarily exceeds its capacity.

Tertiary constraint. The third physical constraint identified during the scheduling process.

Theory of Constraints (TOC). A management philosophy developed by Dr. Eliyahu M. Goldratt that be viewed as three separate but interrelated areas: logistics, performance measurements and logical thinking. Logistics includes drum-buffer-rope scheduling, buffer management, and VAT analysis. Performance measurement includes Throughput, Inventory and Operating Expense and the five focusing steps. Thinking process tools are important in identifying the root problem (current reality tree), identifying and expanding win–win solutions (evaporating clouds and future reality tree), and developing implementation plans (prerequisite tree and transition tree) (APICS).

Throughput. One of the key measurements used to manage a TOC company. The rate at that the system generates money through sales.

Throughput accounting. The use of the three basic measurements of Throughput, Inventory and Operating Expense to manage the financial/ accounting aspects of the company and in making decisions.

Throughput chain. A unique chain of resources, processes or products connected by the generation of Throughput. (Also referred to as **Throughput channel**).

Throughput dollar days (T$D). A measurement used to minimize those activities that threaten the constraint. It is represented by the formula Throughput dollar value × Number of days late.

Throughput justification. The process of justifying expenditures based on their impact to Throughput.

Time line. The time on the planning horizon.

Time rods. A time mechanism used to separate/protect orders (batches) on schedules for different resources.

Time zero. The current date and time.

T-Plant. T-Plants are characterized by a relatively low number of common raw materials and component parts optioned into a large number of end items.

Tracking zone. Zone 2 of the buffer, requiring that orders not having been received at the buffer origin before a specific time be located and tracked for possible problems.

Transfer batch size. The number of parts moved from one resource to the next during production.

Transition tree (TT). Used to define those actions necessary to achieve the goal.

TQM II. The application of the focusing mechanisms and tools of the Theory of Constraints to the tools and methodologies provided by Total Quality Management.

Undesirable effect (UDE). In the current reality tree of the TOC thinking process, identifies those effects in the environment that are unwanted.

VAT analysis. The analysis of how resources perform based on the way in which they interface with each other.

V-Plant. A plant characterized by constantly diverging operations with a small number of raw materials items being converted into a large number of end items.

Zone 1. See **Expedite zone.**

Zone 2. See **Tracking zone.**

Zone 3. That portion of the buffer that does not include the expedite or tracking zone.

Zone profile. The percentage of on time delivery for zones 1 and 2 of the buffer.

Bibliography

American Production and Inventory Control Society. *Production and Inventory Management Journal 32*(1): 7 (1991).

Doty, Leonard A. *Statistical Process Control.* Industrial Press, New York (1991).

Goldratt, Eliyahu M. *The Haystack Syndrome.* North River Press, Croton-on-Hudson, NY (1990).

Goldratt, Eliyahu M. *It's Not Luck.* North River Press, Great Barrington, MA (1995).

Goldratt, Eliyahu M. *The Theory of Constraints: A Systems Approach to Continuous Improvement.* Delmar, Albany, NY (1995).

Goldratt, Eliyahu M., and Cox, Jeff. *The Goal.* North River Press, Great Barrington, MA (1992).

Goldratt, Eliyahu M., and Fox, Robert F. *The Race.* North River Press, Croton-on-Hudson, NY (1986).

Hall, Robert W. *Zero Inventories.* Dow Jones Irwin, Homewood, IL (1983).

Hauser, John R., and Clausing, Don. The house of quality. *Harvard Business Review 66*(3): (1988).

Imai, Masaki. *Kaizen: The Key to Japan's Competitiveness*, Random House, New York (1986).

Ishikawa, Kaoru. *What Is Quality Control?* Prentice Hall, Inc., Englewood Cliffs, NJ (1985).

Japan Management Association. *Kanban: Just in Time at Toyota.* Productivity Press, Cambridge, MA (1986).

Juran, J. M. *Juran's Quality Control Handbook,* Fourth Edition. McGraw-Hill, New York (1988).

Kiemele, Mark J., and Schmidt, Stephen R. *Basic Statistics: Tools for Continuous Improvement.* Air Academy Press, Colorado Springs, CO (1991).

King, Bob. *Better Designs in Half the Time: Implementing (QFD) Quality Function Deployment in America.* Goal/QPC, Methuen, MA (1987).

Lockamy, Archie, and Cox, James F. *Re-Engineering Perfomance Measurements*. Irwin Professional Publishing, New York (1994).

Mizuno, Shigeru. *Management for Quality Improvement: The 7 New Tools*, Productivity Press, Cambridge, MA (1988).

Nakajima, Seiichi. *Introduction to Total Productive Maintenance*. Productivity Press, Cambridge, MA (1988).

Noreen, Eric, Smith, Debra, and Mackey, James T. *The Theory of Constraints and Its Implications for Management Accounting*. North River Press, Great Barrington, MA (1995).

Oakland, John S. *Total Quality Management*. Nichols Publishing, New York (1990).

Orlicky, Joseph. *Material Requirements Planning*. McGraw-Hill, New York (1974).

Osburn, Moran, Musselwhite, and Zenger. *Self-Directed Work Teams: The New American Challenge*. Business One Irwin, Homewood, IL (1990).

Ryan, Thomas P. *Statistical Methods for Quality Improvement*. John Wiley and Sons, NY (1989).

Scherkenbach, William W. *Deming's Road to Continual Improvement*. SPC Press, Knoxvill, TN (1991).

Shingo, Shegeo. *Zero Quality Control: Source Inspection and the Poka Yoke System*. Productivity Press, Cambridge, MA (1986).

Stein, Robert E. Beyond statistical process control, production and inventory. *Management Journal 32*(1): (1991).

Stein, Robert E. *The Next Phase of Total Quality Management: TQM II and the Focus on Profitability*. Marcel Dekker, New York (1994).

Stein, Robert E. *Re-Engineering the Manufacturing System: Applying the Theory of Constraints*. Marcel Dekker, New York (1996).

Suzaki, Kiyoshi. *The New Manufacturing Challenge*. Free Press, New York (1987).

Umble, Michael M., and Srikanth, M. L. *Synchronous Manufacturing: Principles of World Class Excellence*. Southwestern Publishing Co., Cincinatti, OH (1990).

Vollman, Thomas E., Berry, William L., and Whybark, D. Clay. *Manufacturing Planning and Control Systems*. Dow Jones Irwin, Homewood, IL (1988).

Index